D0341071

Mr. Gatling's Terrible Marvel

MR. GATLING'S TERRIBLE MARVEL

The Gun That Changed Everything and the Misunderstood Genius Who Invented It

JULIA KELLER

VIKING

VIKING
Published by the Penguin Group
Penguin Group (USA) Inc., 375 Hudson Street,
New York, New York 10014, U.S.A.
Penguin Group (Canada), 90 Eglinton Avenue East, Suite 700, Toronto,
Ontario, Canada M4P 2Y3 (a division of Pearson Penguin Canada Inc.)
Penguin Books Ltd, 80 Strand, London WC2R 0RL, England
Penguin Ireland, 25 St. Stephen's Green, Dublin 2, Ireland
(a division of Penguin Books Ltd)
Penguin Books Australia Ltd, 250 Camberwell Road, Camberwell,
Victoria 3124, Australia (a division of Pearson Australia Group Pty Ltd)
Penguin Books India Pvt Ltd, 11 Community Centre,
Panchsheel Park, New Delhi–110 017, India
Penguin Group (NZ), 67 Apollo Drive, Rosedale, North Shore 0632,
New Zealand (a division of Pearson New Zealand Ltd)
Penguin Books (South Africa) (Pty) Ltd, 24 Sturdee Avenue,
Rosebank, Johannesburg 2196, South Africa

Penguin Books Ltd, Registered Offices: 80 Strand, London WC2R 0RL, England

First published in 2008 by Viking Penguin, a member of Penguin Group (USA) Inc.

1 3 5 7 9 10 8 6 4 2

LIBRARY OF CONGRESS CATALOGING IN PUBLICATION DATA
Keller, Julia.
Mr. Gatling's terrible marvel : the gun that changed everything
and the misunderstood genius who invented it / Julia Keller.
p. cm.
Includes bibliographical references and index.
ISBN 978-0-670-01894-9
1. Gatling, Richard Jordan, 1818–1903. 2. Gatling guns. 3. Inventors—
United States—Biography. 4. Gatling, James Henry, 1816–1879. I. Title.
UF620.G3K45 2008
623.4'424—dc22
[B] 2007040438

Printed in the United States of America

Set in Berthold Baskerville Book Designed by Francesca Belanger

To the memory of my father

James Richard Keller (1931–1984)

and

to Susan K. Phillips

CONTENTS

I have read a fiery gospel writ in burnished rows of steel.

—Julia Ward Howe,
"Battle Hymn of the Republic" (1862)

They showed us the new battery gun on wheels—the Gatling gun, or rather, it is a cluster of six to ten savage tubes that carry great conical pellets of lead, with unerring accuracy, a distance of two and a half miles. It feeds itself with cartridges, and you work it with a crank like a hand organ; you can fire it faster than four men can count. When fired rapidly, the reports blend together like the clattering of a watchman's rattle. It can be discharged four hundred times a minute! I liked it very much.

—Mark Twain (1868)

"It's the Gatlings, men, our Gatlings!"

—Theodore Roosevelt (1899)

Introduction

The bravest men can do nothing without guns.

—Rommel

Did Richard Jordan Gatling know he had changed the world? Was there a moment when it became clear to him?

Others had tried to create what he created. They tried for centuries. They failed. Their inventions blew up, or were wildly inaccurate, or jammed too easily. His, however, did the trick. It functioned beautifully, and for the first time in history, death was automatic. Death could be reliably doled out in sweeps and clusters, in reeling multiples, instead of one by one. Hence a world that had been moving steadily toward an enlightened recognition of the significance of even a single life, progressing toward a thoughtful and humane acknowledgment of the uniqueness of the individual, suddenly was forced to contend with the appalling opposite, with an ugly new truth: People could be erased with the simple pivot of a gun barrel, with the calculated fury of a devastating weapon. Death was mechanized. Human beings were interchangeable, just as were the parts in other new machines, machines that functioned in humming lockstep as part of the brightly efficient new manufacturing techniques that increasingly defined the age.

Civilization thus was irrevocably altered. Nothing—neither warfare nor diplomacy nor science nor business nor technology nor literature nor art nor theology—would ever be the same. Could Richard Gatling, no matter what his motives were, have had any idea what he had unleashed on the world?

Ah, you say, but if *he* hadn't done it, then someone else would have invented the first successful machine gun. Isn't that right? By the middle of the nineteenth century, the world surely was ready for such a weapon. It was poised to leave behind the arcane intimacy of old-fashioned ways of battle, of flashing knives and clumsy muskets, of enemies falling together in a death-grip that mimicked a lovers' embrace. It was prepared for war to become a thing of detached and distant anonymity, of slaughter on a vast scale. Someone else would have invented an effective machine gun—there were many men who hoped to do just that—but Gatling did it first. As Robert Oppenheimer wrote of Albert Einstein, "The discovery of quanta would surely have come one way or another, but he discovered them."

Lethal weapons complicate America's conception of itself. Long before domestic gun control became a contentious issue in presidential politics, American citizens were already arguing about the use of military force: When is it appropriate to go to war? Does might make right? The debate over armaments and ethics has long been so intense, so spirited, that the fascinating business and cultural history of firearms, and their central place in the American system of manufacture in the early nineteenth century, often is overlooked. It is surely true that to possess a gun is "to hold a piece of death in your hands," in the starkly resonant phrase of contemporary crime writer George Pelecanos. It is not, however, the only truth worth knowing about guns.

Guns were among the first fruits of the Industrial Revolution in the United States, among the first commercial products to benefit from the amazing new concept known as the interchangeability of parts. Yet when historians and cultural commentators discuss the nineteenth century, they readily tick off familiar categories such as transportation and communication. They rhapsodize about railroads and steamships and telegraphs—and often neglect to mention one of the chief transformational elements of that crucial century, which is a gun.

Not just any gun. The Gatling gun.

Most people know what a Gatling gun is. They know what it looks like and how the original version operated: the bundled barrels, the hand crank. Yet Richard Gatling himself remains trapped in the shadows. Famous in his own time, he is largely forgotten in ours. That relative obscurity, measured against the size of Gatling's achievement, is compelling evidence of the odd, murky status of armaments in American life. To this day, a slang term for a firearm is a "gat." His invention is a household name—"Gatling gun" still is a common metaphor for anything that comes swiftly, unchecked, pell-mell—but the man who created it is scarcely common at all. The popular television series *Gilmore Girls*, with its witty, fast-talking characters, sometimes is described as featuring "Gatling-gun dialogue." Yet when people are asked about the origin of the gun's name, they tend to shrug and look around and finally speculate that . . . well . . . okay, maybe it was invented in Gatlinburg, Tennessee. (That city is named for a storekeeper named Radford Gatlin—with no "g" on the end—who lived there briefly in the 1850s and has no connection to Richard Gatling's North Carolina family or to the Gatling gun.) The existence of Richard Jordan Gatling—a key inventive figure of his time, at one point in American history as famous and celebrated as a Henry Ford or a Bill Gates were and are in their times—is unknown to most of his countrymen and countrywomen. He is a bit of a black hole in history, the same history his gun utterly transformed.

That is because in America, guns are never just guns. Guns are singular. The nation's ambivalence about weaponry surely is complicit in the long habit of overlooking the Gatling gun as a cultural symbol, as a superb lens through which to look back at the nineteenth century, and it surely accounts for a good part of Richard Gatling's obscurity. The intensity of that ambivalence is unique to the United States because of the extraordinary nature of what the nation likes to think it represents: triumph not through superiority of arms, but through superiority of ideas. And peace-loving people are reluctant to acknowledge the signal importance—to history, to politics, to business, to philosophy, to culture—of instruments of death. Americans are uncomfortable with the notion of granting guns such pride of place.

The national ambivalence over armed might can be traced back to the beginnings of the republic. In the early 1800s, when the British navy was attacking American ships and seemed to be spoiling for a fight, President Jefferson hesitated to resort to a military response—but not because he was afraid his country might lose. He was afraid it might win. And that, in turn, would pump up the power of the central government, to which Jefferson was philosophically opposed. "War, which every other nation in history had looked upon as the first duty of a state, was in America a subject for dread, not so much because of possible defeat as of probable success," was how Henry Adams phrased it in his history of the early United States.

When it comes to guns, America has a historical blind spot. Its citizens revere the still-life image of a gaunt, moody Abraham Lincoln agonizing over the seesawing fortunes of the Civil War. But they block out the reality of the vigorous, energized Lincoln on one of his frequent excursions to the Washington Navy Yard to test yet another new gun. Lincoln was vitally interested in weapons, especially experimental ones. As an inventor himself—Lincoln is the only president to hold a patent, having been issued U.S. Patent No. 6,469 in 1849 for an inflatable tube that could be rigged to the bottom of steamboats to help haul them over sandbars—he was fascinated by the application of technology to armaments. Few Americans, though, can readily or comfortably envision a lively, gun-toting Lincoln. They are far more at home with the portrait of the thoughtful statesman, of the lean-cheeked orator intoning "Fourscore" with solemn gravity.

The Gatling gun changed everything. It was the world's first machine gun that actually worked—as opposed to the many whose designs looked dandy on paper but which never could be made to operate safely and reliably. It was a pivot point not only because of its innovative craftsmanship and brutal efficiency, which in its original 1862 incarnation sent some two hundred shots per minute crashing out of six barrels, but also because of how that efficiency changed humanity's conception of itself. For the first time in history, you could kill an enemy en masse. At a time when the idea of the individual was rising to unprecedented cultural prominence, the machine gun shoved that individual right back down into the undifferentiated murk, back into a bloody blur, back into nothing but "the sickle shape of the fallen" after "the machine gun had raked in an arc," as Sebastian Barry so vividly describes World War I casualties in his novel *A Long Long Way*.

Killing no longer meant a one-on-one contest. You could kill without looking your opponent in the eye. You could kill without even knowing how many people you had killed. The Industrial Revolution had changed manufacturing from a matter of craftsmanship by individual artisans to a matter of the assembly-line labor of anonymous factory workers, and then the Gatling gun and its deadly spawn—such as the AK-47—came along and turned the heretofore intensely personal work of armed combat into the impersonal work of machines.

Perversely, these new possibilities for mechanized destruction, enabled by the Gatling gun, arose at a time when the cultural world gave every indication of moving in the opposite direction: away from the mass, away from the smear of sameness, and toward a realization of the intrinsic value and unique contributions of each individual. More than three million men died in the Napoleonic Wars, their bodies "shoveled into mass graves" and "their remains even re-used as agricultural fertilizer." But by the late nineteenth and early twentieth centuries, nations began to individualize their war dead in monuments that listed row after row after row of specific names. After World War I, "the Western European belligerents quickly established that all dead soldiers of whatever rank would be buried in special cemeteries." The Romantic movement in literature, which blossomed across the nineteenth century in a variety of genres and guises, helped inculcate the fixation on individual selves. But could such an idea also have arisen to some degree to counter just those dehumanizing effects of the innovative weaponry? When people realized how meaningless a single human life could be—and nothing brought the point home more swiftly and definitively than the sight of a Gatling gun, its perfect circle of slender barrels gleaming smartly in the fresh dawn of a world newly besotted by technology—they found themselves anxiously groping for another kind of idea. They yearned to believe that individuals do matter, that discrete lives do possess a kind of poetry and significance belied by the fact that they can be rinsed away with one casual swipe of a Gatling gun. A terrible paradox was born.

The Gatling gun resonates for other reasons, too. Its creation reveals a great deal about American inventiveness and entrepreneurial drive in the last half of the nineteenth century. There had been many designs and prototypes for machine guns before Gatling tried his hand at one—Leonardo da Vinci was among those who dreamed up a multiple-firing weapon—yet unlike the others, Gatling's worked. It worked spectacularly well. And Gatling realized, better than many of his fellow inventors, that advertising and sales in a newly interconnected world marketplace could be almost as important as the quality of the invention itself. That is why the Gatling gun is a special part of the story of America's astonishingly swift climb to global economic preeminence.

Why did he do it? For all of Richard Jordan Gatling's cool-headed technical finesse and businessman's brio, he actually came up with his gun, he claimed, for the most tender-hearted of reasons: as a way of saving lives. "It occurred to me," he wrote to a friend in 1877, "that if I could invent a machine—a gun—which could by rapidity of fire, enable one man to do as much battle duty as a hundred, that it would, to a great extent, supersede the necessity of large armies, and consequently, exposure to battle and disease be greatly diminished." As disingenuous and self-serving as that sentiment sounds, it ended up being quite correct: Innovations in arms steadily reduced the relative lethality of battles (not to mention the cost of waging war) throughout the twentieth century.

The same justification was embraced by many of the scientists who hot-wired America's atomic arsenal less than a century after Gatling created his gun. Horrifically destructive as their handiwork was, still they believed they were toiling for what would ultimately prove to be the greater good. You can chalk this up to a painfully obtuse naïveté, or you can see it as the very human and surely admirable impulse to wed the best of intentions to the latest in scientific progress, and to believe that this time—*this time*—things really will be different. Separated by time and technological advances, Gatling and Oppenheimer were brothers under the skin.

The Gatling gun foreshadows another paradox, too, the paradox of American success itself, the fact that a nation established on a radiant ideal about human liberty owes a large portion of its leadership position in the world not to that ideal but to military strength, to the subduing of enemies with firepower. The tension between America's sunny conception of itself and the reality of its role in the modern world—the role of bully and enforcer—reverberates down to the present day. Military might is an essential component of the American dream; it is a component, however, that disturbs many Americans.

Early on, some military leaders rejected the Gatling gun because it seemed to snatch from warfare the ideal of individual valor that gave battle its patina of nobility and self-sacrifice. So did some in the rank and file. "One couldn't pin a medal on a weapon" was the arch observation of historian John Ellis. The Gatling gun somehow seemed unfair and unsporting, an "unmilitary gimmick." It was like hiding a rock in a snowball. Because even within something as chaotic and terrible as war, there were, some believed, aspects to be savored. There were standards and rituals and clear outcomes. There was a grim, simple logic. War reduced life's bewildering complexities to a series of stripped-down dichotomies: victory or defeat, good or bad, life or death, honor or disgrace. War clarified the world. It was a chance to test one's sincerity and grit. And then along came a piece of hot metal that could, by creating "a crisscrossed lattice of death," as a stunned British soldier described the effects of machine-gun fire in World War I, rob war of its larger symbolic meaning, of the opportunities it offered for personal heroism.

Even before World War I, the Gatling gun had become a symbol—but a symbol of something quite different from honor and glory. In the 1870s, 1880s, and 1890s, in both the United States and Great Britain, the sight of the multiple barrels of a Gatling gun worked to such sinister effect that it became an emblem—both as a threat and as a reality—of government power and institutional authority. In the United States, the deployment of Gatling guns by strike breakers at a labor riot, or by soldiers in the West against the native population, or by policemen at the barest hint of the formation of a mob, was a chilling demonstration of the iron rule of the dominant classes. Gatling guns were routinely wheeled out to make a highly theatrical but inarguable point about who was really in charge. "In the early stages of an acute outbreak of anarchy a Gatling gun, or if the case be severe, two, is the sovereign remedy," lectured an 1886 editorial in the *New York Times*, in the wake of the deadly labor unrest in Chicago's Haymarket Square. In the battle of the ordinary working stiff versus the wealthy and powerful, the Gatling gun was very much regarded as the mascot of the latter.

And it wasn't just the fact that Gatling guns were firearms, which have always been artful persuaders. There was something especially dominating and dehumanizing about the work of a Gatling gun, about a set of barrels fixed around an axis with the pitiless exactitude of any other machine—except that with this machine, the idea wasn't creation, but destruction. When Albert Parsons, a chief agitator at the Haymarket riot, faced the crowd on that tragic day, he cried, "Do you know that the military is under arms, and a Gatling gun is ready to mow you down?" *Mow you down:* His language choice is crucial. Gatling guns didn't just kill; they casually reduced human beings to a row of bland targets. Thus a machine with its interchangeable parts turned people, too, into just so many interchangeable parts, parts without separate identities or unique souls. The first working machine gun wasn't a transformative technology simply because it caused death in great quantities; its unprecedented influence was based as well upon the autonomy it stripped from its victims in mere seconds. Men, women, and children were like stalks of wheat beneath a scythe. *Mow you down.* The world had come a long way from the elaborate ritual and personal grudge match of the formal duel, or even of the spontaneous gun battle on the streets of a scruffy Western town, where individuality mattered.

During this same period—the dwindling decades of the nineteenth century—in Great Britain, those in charge bolstered the Gatling gun's image as an adjunct to oppression and empire, as the handmaiden to racism. Gatling guns were freely employed by British troops in faraway places against opponents who did not look like them—that is, against the indigenous populations of Africa—with little hesitation. There was a sense that the impersonal slaughter visited upon these enemies was acceptable, whereas the same kind of mechanized malevolence would be wholly inappropriate against white Europeans. It wouldn't be proper or seemly to use against "civilized" people.

The Gatling gun thus became the indispensable tool of the racism implicit in Great Britain's imperialist adventures. Cutting down row upon row of "natives" was fine, but such an ignominious fate would have been unthinkable for foes who looked like *us*. The old niceties, the courtly traditions of warfare—a motif found in what historian Paddy Griffith calls the "aristocratic 'lace wars' of the eighteenth century"—still held force, but only when the opponent was sufficiently "evolved." Otherwise, the swift erasure of an entire swath of human souls with one sweep of a Gatling gun was perfectly acceptable.

Sir Henry Newbolt's familiar 1897 poem chronicles just such an encounter on African soil. British troops—smartly uniformed graduates of elite universities, trained to display a square-chinned social correctness and imbued with a sense of chivalry and fair play, topped off with boundless reservoirs of fortitude and dash—face a desperate situation:

> The sand of the desert is sodden red, —
> Red with the wreck of a square that broke; —

> The Gatling's jammed and the Colonel dead,
> And the regiment blind with dust and smoke.
> The river of death has brimmed his banks,
> And England's far, and Honour a name,
> But the voice of a schoolboy rallies the ranks:
> "Play up! play up! And, play the game!"

The troops will rally, the Gatling will be ungummed, and what the British consider the natural order of things—the hierarchy that places the white male European at the top of the heap—will be reasserted, thanks to a magnificent new weapon. Of course, one would never presume to use such a barbaric instrument against one's own kind, would one?

The Gatling gun is a weapon of death, but its story is not altogether grim. For it is also the story of a nation on the rise and of a man who, by inventing a new kind of machine, helped propel it in that upward trajectory. It is the story of a country just at the moment when its destiny begins to stir, and of an individual whose career was hitched to that amazing creative and economic boom. It is the story of one genius who helped push America to the top, a man of decency and vision and ambition, a man who held dozens of patents for a variety of life-enhancing gadgets but who died disillusioned, his name attached in the popular mind not to plows or bicycles or flush toilets or dry-cleaning machines, all of which he improved, but to a gun. A utilitarian device whose use came down to a chilling simplicity: death.

Gatling illustrated "that American poetry of vivid purpose," as William Dean Howells put it in his great nineteenth-century novel, *The Rise of Silas Lapham,* a story whose title character seems a lot like Gatling: smart, hardworking, morally upright. Gatling was part of a generation of Americans who made the modern world, who took an overgrown, half-formed hunk of wilderness still broken and dazed by the Civil War and shoved it into the next century by force of will and wizardry of invention. And they did it while insisting on a personal moral code, on a philosophy that George Eliot, in another illustrious nineteenth-century novel—*Middlemarch* (1871)—called "life beyond self." It is that very earnestness, that sense of fealty to something larger and more abstract than their own success—although great success they certainly had—that bewitches the story of America's nineteenth-century industrial and financial titans. Just being rich didn't make them interesting; what made them interesting is the fact that their wealth was haunted by obligation. Andrew Carnegie built libraries, John D. Rockefeller established universities, J. P. Morgan funded museums. As a historian of the period has noted, those possessing the great fortunes of the day eagerly "ushered in a regime of philanthropy on a monumental scale."

Like the fictional Lapham, the self-made millionaire who loses his money through bad breaks and through an unwillingness to be anything less than honest in his business dealings, Gatling became a very rich man and then a poor one. But as we shall see, he never seemed to lose that quintessential American optimism, the sense that a second fortune is always just beyond the horizon, the conviction that there is always something new to make and someone else to sell it to.

The nineteenth century was a heroic intellectual age, an age of great writers and theorists whose ideas supplied a vital source of energy in America's uncannily swift climb from bashful newcomer in the world community to confident leader by early in the next century. But those people are only part of the story.

You can read about the intellectual underpinning of the nineteenth century and admire those marvelously inspirational thinkers, but you might find yourself asking this question: What did they *do*? How did these men with fierce opinions but soft hands ever create the twentieth-century colossus known as the United States? The Emersons and Thoreaus and Whitmans and Jameses talked, talked, talked, and wrote, wrote, wrote, but what did they *make*? Could it be that the great mechanical genius and entrepreneurial drive of this era helped create the vibrant intellectual achievement—instead of the other way around?

Gatling made many things, from guns to plows to propellers. He had an engineer's mind for the nuances of how things worked and a storekeeper's eye for the dollar. And while eras may be explored through their politics and their literature and their philosophies, they also can be understood through their inventions and their businesses, through the men and women who put their reputations and their bank accounts on the line day after arduous day, night after sleepless night, in pursuit of a dream.

America may have formally gotten under way in the 1700s, but the true dawn of its special character is right here—here in the long, famished sigh following the Civil War. It could have been a bleak time, a resigned time. It was not. Instead, it was a time of progress, a time when a handful of risk-taking, farsighted businessmen awakened each morning determined to get things done, to secure their own fortunes by the force of their ambitions and by the audacity of their deeds.

This was when America became America: a country renowned for its vigor, its riches, its technological innovation, its cultural upwelling, its indomitable optimism. America had more than doubled itself, noted Henry Adams, and while the nation's population in the first half of the nineteenth century indeed increased by almost 450 percent, Adams didn't mean just people and property. He also meant that ineffable sense of national possibility, a flying wedge of such relentless forward momentum that not even a violent internal quarrel could throw it off stride for long. America was a gathering force. "It was during this period," wrote historian William Cronon, "that much of the world we Americans now inhabit was created: the great cities that house so many of us, the remarkably fertile farmlands that feed us, the transportation linkages that tie our nation together, the market institutions that help define our relationships to each other."

What does all of this have to do with a gun? Everything. The Gatling gun changed the way wars are fought; that much is obvious. Yet it also changed the way people thought about their lives, and it changed how they came to live those lives. It changed how America saw itself, and how other nations saw America. And in the kind of unintentional irony that also seems deeply American, the man responsible for the Gatling gun intended it to be a tool of peace. Its brutal spit-spot efficiency would, he hoped, persuade nations of the waste and folly of war.

The world is always changing, of course, but the breadth and velocity of the changes at this time were remarkable. Charles Darwin's ideas were challenging scientists to look at the history of living things in an entirely new way, at the means by which species alter over time: Some varieties flourish, others perish. The population was linked—by railroads, by telegraph and telephone, by national publications—into a mass market, and this new market was shaking up the economic system. Ballooning numbers of people were seeking the same few goals: money, land, higher standards of living. There was a sweeping and monumental shift in emphasis from the one to the many. Darwin's theories—or at least the way those theories were presented to the public—made war appear inevitable, as more and more people would fight for what seemed to be fewer and fewer resources. Darwinian ideas "promoted the notion that war was part of the natural order of things," writes historian Charles Townshend. "European society on the eve of the First World War was becoming increasingly bellicose."

The Gatling gun, too, forced a change in perspective from the individual to the aggregate. It encapsulates one of the great intellectual debates of the nineteenth century: whether we live in a handcrafted world, a one-at-a-time world created by a prim and attentive god, or in a mass-produced world, a world that emerges through the randomness of natural selection and a blind, machinelike pulse. An intimate world, a world made on purpose—or an inadvertent world, a universe of happenstance, the accidental residue of time and chance and wind direction. It is a world in which each life is sacred and irreplaceable. Or it is not.

The Gatling gun may have been the product of a single human will, but it seemed, in its all-purpose sweep of destructiveness, to erase the possibility of individual wills. That mystery haunts the latter part of the nineteenth century. It seeps into the twentieth. It still matters in the twenty-first.

So this is the tale of a gun, and of the man who made it, and of the world he both inhabited and, through his skill and his pride and his passion and perhaps even his obtuseness, changed forever. Richard Gatling, like the United States, is a contradiction. He was naïve and ruthless. He was soft-hearted and calculating. He was a decent and honorable man, but he was also a CEO with an eye fixed on the bottom line. In a letter to one of his sales representatives, dated June 18, 1872, Gatling asks about an ailing friend and former business associate—"I regret to learn that Mr. Talbott is no better. I hope he may yet recover"—and then switches, a few lines later, to a rather different worry: "Try to avoid selling sample guns; by doing so, we may delay chances to get larger orders." The concerned friend, the focused businessman: He was both, and sometimes he was both within the same paragraph. Thus was Gatling a symbol of his country's restless, bifurcated soul. One half glowed with idealistic fervor. The other half glowered with a burgeoning appetite for conquest and domination. And his gun put the world on notice: America had arrived.

CHAPTER ONE

COLD BEAUTY

Nothing except a battle lost can be half so melancholy as a battle won.

—Duke of Wellington

Oliver Hazard Perry Throck Morton was a man of immensities. He had an extra-long name, an extra-large body, grand ambitions—and a big problem: the Civil War.

Things weren't going at all well for President Lincoln and the Union side. Morton, who had settled his ample rump into the Indiana governor's chair a scant three months before the war began on April 12, 1861, was firm in his loyalty to Lincoln, a fellow Republican. Morton was on board from the beginning. With the fires still smoldering at Fort Sumter, the president had requested six regiments from Indiana. Morton's jaunty reply: Take an even dozen.

Yet the swift and sure defeat of the rebels, an outcome predicted with such lavish confidence by so many, now seemed less likely with each passing day, with each dismaying report from the front lines.

"To His Excellency Governor Morton," read a typical report, this one dated April 17, 1862. "Sir: The following are the names of wounded soldiers, now in the Hospital . . . ," and what followed, in the flourishing script of the governor's correspondent, was a somber and dreary list: "James B. Mullen . . . compound fracture of right thigh bone" and "Henry Reddington . . . left arm amputated" and "William Maloney . . . wounded in right arm" and "Benjamin

Smith . . . gun shot wound in leg" and "Harvey Harlow . . . shot through the breast."

On and on it went, a steady drumbeat of loss, of pain and death inflicted on Indiana's volunteer regiments. The grim lists piled up on Morton's desk like field notes from hell.

Morton, more than most governors, had staked his political reputation on his support for Lincoln. Each time the president requested more volunteers, Morton rallied his state's citizens to the cause. The portly, pugnacious governor vehemently opposed the idea of a draft. Anteing up a reliable supply of volunteers would, Morton hoped, head off attempts to impose conscription in his state.

But who had ever dreamed it would come to this? Who could have envisioned this sad and bloody tangle that inflicted fresh misery with each day's news?

In the beginning, almost no one had foreseen a long war. Few had imagined that the rebellion would turn into a ghastly, confusing stalemate, into an appallingly lethal struggle requiring a constant supply of young bodies.

William Seward, Lincoln's secretary of state, had breezily predicted that the war would last a mere ninety days. Editorial writers at the *Chicago Tribune* had boasted that it would all be history in two to three months, because "Illinois can whip the South by herself." Not to be outdone in the prognostication department, the august editors at the *New York Times* had tilted their heads thoughtfully, put finger to chin, and estimated that the war would last thirty days, tops.

Then came July 21, 1861, and the Southern victory at Bull Run, and the gruesome dawning of a terrible new truth: The war would be neither easy nor quick. Old assumptions crumbled. Expectations were swept away. Proud certainties now seemed excruciatingly naïve.

While the North had unquestioned superiority in numbers, in money, in manufacturing capability and technical finesse, wars have a funny way of circumventing the careful logic of their planners. As

U. S. Grant would note in his memoirs, the South initially received a boost from its upstart status. Its officers were gallant and able men with whom Grant had attended West Point, with whom he had served in the Mexican War—experienced men, that is, men of discipline and courage and drive. The Confederate cause "had from thirty to forty per cent of the educated soldiers of the Nation," Grant wrote. "They had no standing army and, consequently, these trained soldiers had to find employment with the troops of their own States. In this way what there was of military education and training was distributed throughout their whole army. The whole loaf was leavened."

Perversely, the North's very supremacy in firearms ended up working against it. Because so many early battles unexpectedly went the Southerners' way, a great many Union weapons—chiefly Model 1861 Springfield and Pattern 1853 Enfield muskets—were confiscated and later employed by rebel forces. The North had been sluggish about ordering guns from abroad; the firearms shortage had grown desperately acute. "Twenty-four hundred men in camp and less than half of them armed," Morton fumed in a letter to Secretary of War Cameron. "Why has there been such delay in sending arms? . . . Not a pound of powder or a single ball sent us, or any sort of equipment."

Battles that should have been won were lost, or simply unraveled into gory toss-ups with no clear victor at all. Advantages that might have been pressed were squandered. And still the war went on, chewing up volunteers and infuriating governors such as Morton, who refused to be passive in the face of what they saw as flagrant mismanagement.

Morton knew about guns, and he knew that the old smoothbore muskets with which the army was mainly equipping his men were ludicrously inadequate. Early in the war the incensed governor whipped off a scathing letter to Lieutenant Colonel James W. Ripley, the army's chief of ordnance: "It is the opinion of all military men here that it would be little better than murder to send troops into battle with such arms as are a large majority of those muskets," he thundered. Getting no satisfaction from Ripley, Morton griped to the Secretary of War that he had faced "embarrassment in transacting business with General Ripley from the beginning of the war." The governor, who never minded a bit of name-calling in service to a noble cause, added that an assistant to Ripley was "superannuated, fretful, and slow, and not very much superior to General Ripley as a businessman."

Morton was the governor of an important state, so his verbal volleys could not be ignored. Six months into the war, Lincoln replied in a wire that seems to ache with weary candor: "I wish you to believe (as we certainly believe of you) that we are doing the very best we can. You do not receive arms from us as fast as you need them; but it is because we have not near enough to meet all the pressing demands. . . . We have great hope that our own supply will be ample before long, so that you and all others can have as many as you need."

No matter how much they smarted over the lack of weapons, though, Morton and other loyal governors would not refuse Lincoln's call for more men, and then more, and again—more. In the first few months after Fort Sumter, the Union army jumped from 16,000 to 486,000, on the strength of the president's summons. A visitor to Washington, D.C., shortly after the war's commencement recorded that the air was filled with the cry, "Still they come," as recruits streamed in from across the country.

Months passed, the war dragged on, a beleaguered Lincoln kept asking, and Morton kept obliging.

In July of 1862, after Lincoln's most recent request for three hundred thousand additional volunteers, an abolitionist poet named James Sloan Gibbons wrote a stirring four-stanza poem, an earnest exhortation reflecting the grim necessity for additional sacrifice. The people must send more and more and yet *more* of their sons to fight and die along the brown rivers or on the bleak plains or in the scorched forests of a deadlocked land.

Several composers rigged the poem to music. L. O. Emerson's tune was the one that caught on, and within its singsong rhythm, within its cadenced sentimentality, was the innocent fervor of the sacrifice made by ordinary people:

> We are coming, Father Abraham, three hundred thousand
> more,
> From Mississippi's winding stream and from New England's
> shore;
> We leave our ploughs and workshops, our wives and children
> dear,
> With hearts too full for utterance, with but a silent tear;
> We dare not look behind us, but steadfastly before:
> We are coming, Father Abraham, three hundred thousand more!

It was that spirit—reluctant but resolute—that Morton had tapped again and again in the state he led and loved. He was, in fact, the first Indiana governor to have been born within the state's borders, and that seemed to make a difference: He was a native son encouraging other native sons to join the cause. And many were returning maimed and broken, if they returned at all.

A restless Morton was groping for answers, for something—anything—that would shorten the war, that could stop the steady and depressing arrival of dispatches listing the grim fates of Indiana's brave young men. Morton, who possessed massive faith in his own judgment, never minded bending the rules; he prided himself on finding creative solutions, even if they required flexible ethical standards. He was a gutsy pragmatist, this pudgy former hatter's apprentice who ruled his state with imperial flourish. Other governors, when facing sharp questions from nervous legislators about their conduct of the war, might take pains to get along; other governors might wobble and wheedle and humbly explain themselves. Not Morton. He ordered Republican legislators to just go home, thus thwarting the Democrats in control who, lacking a quorum, could take no actions. Morton then ran the state his way for the next two years and funded his adventures with loans of dubious legality from Indiana businesses and cash from a grateful War Department. One enraged Democrat referred to Morton as a "low-flung demagogue"; the word "lunatic" also was tossed about with some regularity.

Morton shrugged it off. He was not a man to ask permission from anybody. He never hesitated. He pushed ahead, a stout steamroller of purposeful certainty.

That sort of indomitable drive would serve Morton well, even after the crisis of the Civil War had passed. A stroke in 1865 would leave his legs dangling and useless, yet still he was elected to the United States Senate in 1867. He had to be carried to his desk each day. But he was one of President Grant's closest advisers and almost snared the Republican presidential nomination for himself in 1876.

The present crisis, though, was proving to be a ferocious challenge, even for an Oliver P. Morton. The governor known for his brashness, his boldness, his pluck, and his bluster, was flummoxed and frustrated. Thus he was open to new ideas, no matter how far-fetched. He was eager to grasp just about any solution, any scheme to get the thing over with.

And he thought he just might have found it.

Morton had learned that a forty-three-year-old Indianapolis businessman had come up with an intriguing idea for a new kind of gun. A gun that did its job with decisive speed and terrific efficiency. A gun that would outpace all other weapons that came before it. A gun that would persuade anyone who witnessed, even once, its sweeping arc of destructive firepower to give up and go home.

Not that other inventors hadn't made similar claims for other guns. Dreaming up weapons, in fact, was a common hobby during the Civil War, so much so that the War Department found itself having to fend off dozens of proposals for newfangled firearms. As the war went on and on, more and more inventors stepped up with their can't-miss cannons or sure-thing rifles or just-try-me muskets. Among multiple-firing weapons, there was the Vandenburgh volley gun, the Ager coffee-mill gun and the Billinghurst-Requa battery gun.

Yet this gun, Morton thought, might be different. It held unusual promise. Its inventor, Richard Jordan Gatling, was no crackpot eccentric, but a respected and socially connected businessman, married to the daughter of a prominent Indianapolis physician. Gatling's sister-in-law was married to David Wallace, former Indiana governor. And the credentials didn't stop there: Gatling also was a friend and confidant of Benjamin Harrison, and Benjamin Harrison was a good man to have in your corner.

Harrison was an Indianapolis attorney hooked up with a prominent law firm. He was an up-and-comer in the state's Republican party. Harrison, many people speculated, just might be a political name to be reckoned with in the future, maybe at the national level. Who knew? In any case, he seemed to be making all the right moves. In the summer of 1862, at Morton's request, Harrison organized and led a regiment of Indiana volunteers; they would return covered in glory. Little wonder, then, that Harrison's word carried a great deal of weight with Morton. Yet it is also true that if the gun had been a bust, if it had fizzled in its field trials like so many other newfangled weapons, then all the glowing recommendations in the world wouldn't have helped Gatling's cause.

Gatling had already made a fortune by patenting an agricultural device, a wheat drill that wowed farmers and swept prizes at state fairs. Before that, he had studied for a medical degree in Cincinnati, and he didn't mind a bit if people chose to call him "Doctor." (Fellow gunmaker Samuel Colt, too, spent a few seasons with the "Dr." prefix attached securely to his surname to lend his image a certain learned authority, although Colt's justification for the honorific was even skimpier.) Gatling had stature and obvious competence. Morton wasn't going too far out on a limb when he decided to back the dapper inventor.

The Indiana governor, who loved to pepper Lincoln and his staff with forthright recommendations on how best to win the war, fired off a note to the Assistant Secretary of War on December 2, 1862:

> Sir—Allow me to call your attention to the "Gatling Gun," invented by Dr. R. J. Gatling, of this city. I have been present at several trials of this gun, and without considering myself competent to judge certain of its merits, am of the opinion that it is a valuable and useful arm. Dr. Gatling desires to bring it to the notice of your Department, with the view of having it introduced into the Service.
>
> I cheerfully recommend him to you as a gentleman of character and attainments, and worthy in all respects of your kind consideration. Any favour you may be pleased to show him will be duly appreciated.
>
> Very respectfully,
> Your obedient servant,
> O. P. Morton, Governor of Indiana

He was writing on behalf of Gatling, to be sure, and his nifty new weapon. But Morton was also writing on behalf of those who had decided, in the quiet crucible of their own hearts, to heed the words of Gibbons' poem, to yield to its softly insistent invective. It was a song whose lilt and charm were terribly at odds with the actual horror of its topic. A bloody lullaby.

> If you look up all our valleys where the growing harvests
> shine,
> You may see our sturdy farmer boys fast forming into line;
> And children from their mother's knees are pulling at the weeds,
> And learning how to reap and sow against their country's
> needs;
> And a farewell group stands weeping at every cottage door:
> We are coming, Father Abraham, three hundred thousand more!

Morton was writing for those who were going to join the fight, a fight that would, God willing, be decisively shortened by a new machine.

A machine *gun.*

It was a splendid-looking thing, a marvel, a curiosity, a dazzle of shiny metal and sleek design, even in its fledgling state. Six metal barrels were arranged in a tight and intimidating-looking circle and affixed on a narrow platform suspended between two large wagon wheels. Jutting from one side of the device was a hand crank.

To speak of a weapon's aesthetic qualities might seem rather perverse, even preposterous, yet there can be a cold beauty to the most lethal of objects. The man who would, less than a century hence, supervise the creation of the most destructive device in the history of the world, the atomic bomb, acknowledged as much: "I have always thought it was a dreadful weapon," J. Robert Oppenheimer wrote of the bomb to a friend in 1954, although "from a technical point of view it was a sweet and lovely and beautiful job."

Sweet and lovely and beautiful: The description fits as well the gun that Gatling displayed to a curious audience on the streets of Indianapolis in that troubled and uncertain spring of 1862.

When the crank was turned, there was an explosion of noise. A flurry of bullets. A flutter of destructive fury.

What the good people of Indianapolis saw that day was the world's first working machine gun. Its inventor had named it after himself—the alliteration was irresistible, the lyrical bounce and cadence of "Gatling gun" was a phrase with a captivating naturalness to it—and Gatling would, in subsequent decades, market it relentlessly around the globe, with style and savvy and evergreen gumption. Gatling's inspiration, an armament historian later noted, was "the first great American invention."

And it was launched by means of a public event, the same kind of well-organized and well-advertised spectacle that would acquaint the nineteenth century with electric lightbulbs and sewing machines and steam boilers and Juicy Fruit gum and telephones. For this was the age of the great public demonstration. Be it a weapon or a washing machine, the way one drummed up interest and support was to arrange an elaborate public show. When young Samuel Colt wanted to inspire an appetite for his electrically detonated underwater mines—this was before he began manufacturing the pistols that immortalized his name—he staged a spectacular demonstration on the Potomac River in 1844 in front of thousands of people, blowing up a five-hundred-ton schooner with a tremendous explosion and thick cloud of black smoke. There was nothing strange or unnerving in 1862 about inviting one's friends and neighbors and business associates to witness the unveiling of a new gun.

But then again, a gun is not like a door latch or a window shade, both of which were touted at nineteenth-century fairs. A gun is singular. Because you could, if you wanted to, draw a line that started on that spring day in 1862 in Indianapolis and ended up at a moment on

April 25, 1915, when British troops stormed ashore on Cape Helles at Gallipoli. Of those first thousand men, some seven hundred were instantly cut down by machine guns wielded by the waiting Turks. Indeed, some 80 percent of all casualties in World War I, said British Prime Minister David Lloyd George, were caused by machine-gun fire. And while the line you started that day in 1862 would not be entirely straight and unbroken, while it might grow fainter and firmer and fainter again across the intervening years, while it would waver at certain points and dip and double back on itself, still you could do it. You could draw that line.

He had improved plows and propellers; now he would work on weapons. Why?

Later, Gatling would try to explain himself in a letter he wrote to a friend in the summer of 1877:

> In 1861, during the opening events of the war (residing at the time in Indianapolis, Ind.) I witnessed almost daily the departure of troops to the front and the return of the wounded, sick and dead. The most of the latter lost their lives, not in battle, but by sickness and exposure incident to the service. It occurred to me that if I could invent a machine—a gun—which could by its rapidity of fire, enable one man to do as much battle duty as a hundred, that it would, to a great extent, supersede the necessity of large armies, and consequently, exposure to battle and disease be greatly diminished. I thought over the subject and finally this idea took practical form in the invention of the Gatling Gun.

It was not as outlandish a rationale as it might seem to contemporary sensibilities. Guns were different then, not just in design, but in the cultural role they played. Gatling was simply following the familiar moral logic of an 1852 editorial in the *Hartford Daily Times,* written to praise Sam Colt's revolutionary new revolver: "Men of science can do no greater service to humanity than by adding to the efficiency of warlike implements, so that the people and nations may find stronger inducements than naked moral suasion to lead them towards peace." The *Indianapolis Evening Gazette* struck the same note in an 1863 editorial in favor of Gatling's innovation: "[Gatling guns] are so light, so easily handled, and require so few men to work them. . . . Every regiment ought to have at least one of them, and it would be well in some cases if every company had one. . . . Three or four of them in such case would be equal to as many fresh regiments, with not one-tenth the danger of loss of life on our side." The Gatling gun was not only the practical choice; it was, many thoughtful people convinced themselves, the ethical one, too.

Thus the prospect of this Gatling fellow, a quiet, gentle, buttoned-down business leader, a family man, shifting his focus to the crafting of deadly weapons was not as unlikely as it sounded. Few of the men who sparked innovations in armaments were, to modern eyes, plausible picks to have done so: They were ministers and lawyers and farmers and schoolteachers. They were playboys and poets. They were presidents—Abraham Lincoln was fascinated by firearms and kept pressing his countrymen to come up with designs for new ones—and engineers and college professors.

Being intrigued by firearms did not, at this point in history, make you a bloodthirsty warmonger. It made you a passionate dabbler. An earnest student of applied technology. An amateur in an age whose most significant inventions—the steamboat, the locomotive, the repeating pistol—were invented by amateurs.

No less a revered figure in the annals of human creativity than Leonardo da Vinci was vitally interested in guns. His design for a wheel-lock pistol, in which a spring-loaded wheel generated sparks by contacting iron pyrite, was the first weapon to carry its own source of ignition; he came up with this revolutionary improvement sometime around 1500, and it soon spread across Europe. Eli Whitney is a familiar name for having patented the cotton gin, but he also made guns; in 1798, he signed a contract with the United States government to turn out ten thousand muskets in the next two and a half years, to help the country better prepare for a possible war with France. Even those who didn't make guns still were intrigued by them: British poet Percy Bysshe Shelley, fascinated by gunpowder, loved to experiment with that lively, unpredictable mix of potassium nitrate, sulfur, and charcoal, a bit of which, set alight, was able to mimic uncannily the elemental destructive fury of nature, of lightning and thunder. Shelley, to his great delight, once blew up a tree stump with a homemade batch of the stuff. Erasmus Darwin, who would pass on his love of experimentation to his famous grandson Charles, had several scary run-ins with the startling effects of gunpowder. Aaron Burr tried his hand at gunmaking before opting for another kind of fireworks: the political world.

Curiosity about guns was perfectly acceptable, and gunmaking was a thoroughly respectable profession. Rural areas in England and America were dotted with families who tinkered with armaments, with father-and-son tandems who turned out firearms for sale or trade. Guns were things, commodities, objects, necessities, not symbols of evil. They were seen, by and large, as products, not as the touchstones for grim moral reflection that they would become a century hence. Gunmakers were not shunned. Sir Henry Bessemer, whose innovative process for oxidizing steel—the Bessemer method—revolutionized industrial capacity, did something else before working on steel refining: In 1854, he devised a breechloading gun that employed steam to feed and fire it. Vickers Brothers, the important British arms manufacturer, began as a flour mill, then switched to steel before finally settling on armaments in 1888, and few people would have considered the move a handshake with the devil. It was a business decision. William Armstrong, another significant British arms manufacturer, had begun his career making hydraulic machinery; the change to weapons was logical and lucrative, given market forces of the day. Armstrong, in fact, sold guns to both sides in the Civil War. In Germany, Ludwig Loewe's company made sewing machines before it made weapons. In the United States, Ivor Johnson Arms & Cycle Works, founded in 1883, turned out guns, bicycles, and roller skates. The plant that made Winchester rifles also made refrigerators. Eliphalet Remington was a barrel maker before switching to guns.

Dealing in guns, a practice that today carries distinctly sinister overtones, had a different connotation then. During the early days of the Civil War, such unlikely figures as financier J. P. Morgan and landscape architect Frederick Law Olmsted engaged in complicated and quite public deals to procure firearms. Olmsted raised money to arm antislavery factions in Kansas and personally selected "a mountain howitzer together with fifty rounds of canister and shell and, for good measure, added five hand grenades, fifty rockets, and six swords." The similarly low-key, distinctly nonbellicose Morgan bankrolled a plan to retrofit five thousand breechloading carbines for use by Union troops, a scheme that netted Morgan a handsome 9 percent profit and 25 percent commission. While Morgan was criticized, the criticism came from his having made a buck on the backs of those brave lads in blue—not from the fact that guns were the products being bought and sold. Yet Morgan "probably saw the loan in purely commercial terms, as involving some risk and reasonably assured reward," wrote biographer Jean Strouse. "The moral question that outraged those who considered it profiteering most likely did not enter his calculations." And the moral question over dealing in guns was never a factor at all.

Guns were regarded as practical necessities, as useful inventions, much like new modes of transportation and communication. And that was why Gatling's method of introducing a new gun—in public, with families present, with fanfare and pomp and advance advertising—was not at all peculiar in that spring of 1862.

Gatling knew the game. He knew that diffidence and hesitation and delay could doom a new enterprise. He knew that holding back held great peril. If you had your eye on substantial profits and if you believed in what you'd made—be it a better butter churn or a better gun—you'd want to make the world aware of it.

To understand Gatling, you have to understand not just what kind of war this was, but what kind of country this was shaping up to be, and what sort of era was unfolding. The nation and the nineteenth century meshed perfectly. The soul of the century was risk and dare, a willingness not only to accept change but also to work for it, to insist upon it, and to look upon what change had wrought in the world and pronounce it to be God's will. The essence of the new country was the same: It was a mammoth land of "overnight cities" and gold strikes and stunning abundance. And if boldness also happened to bring prosperity, if a plan undertaken for putatively noble reasons also smartened up the bank balance—well, why not? Why shouldn't it? The United States and the1800s. They were made for each other.

So he introduced it the way you'd introduce a frying pan or a headache remedy. The way you'd unveil a superior design for a pepper mill or a kite or a bridle. Even with a war on—or perhaps *because* there was a war on—Gatling had no trouble drawing a crowd. People had to take their fun where they could find it. No matter what else was going on, the nineteenth century was a time of carnivals and exhibitions and contests, of large, expectant gatherings and canny promoters, of excitement and pizzazz and extravaganza, of bands and fireworks. People, for the most part, lived hard, boring, and isolated lives, lives of tedium and want and unvarying physical routine, and the prospect of an afternoon of conviviality—of something *different*, for God's sake—was irresistible. If it got you out of the house, if it got you out of the field or the barn, it was special. Who'd want to miss it?

From the logistically complex world's fairs, which demanded years of fund-raising and construction, to the one-night stands in one-horse towns operated by touring companies such as Buffalo Bill's Wild West and innumerable imitators, this was the century of boldly audacious showmanship: *Watch closely now. This is new!* Steamboats, six-shooters, balloons, telephones, locomotives: No matter what the product was, no matter what the thing did or how well it did it, public gatherings were the perfect means to capitalize on natural human curiosity, on the persistent appetite for the new. Fairs popped up in places such as Nashville and Omaha and Buffalo and Atlanta and San Diego and New Orleans, as well as in larger cities such as Philadelphia and Chicago.

When President Polk visited the National Fair in Washington, D.C., the day after its official opening on May 22, 1846, he praised what he saw as "highly creditable to the genius and skill of our countrymen," but don't be fooled by the careful, tepid talk. The fair was a sensation. How could it not be? People strolling the aisles at such events could get a load of coffee mills, ice cream freezers, sausage cutters, washing machines, shovels, cotton looms, and water pumps. They could gawk appreciatively at household goods and farm tools, at whimsical items and at lethal weapons, all brought together to startle and charm.

Pistols and false teeth: These were the hot-ticket items at the Crystal Palace exhibition in London in 1851. Both were invented by Americans. Also on display was Richard Gatling's seed planter; he hadn't yet turned his attention to guns. The popularity of American-made inventions at the 1851 fair, in fact, was what finally convinced British authorities that they needed to change their patent system—the first update in British patent law in two hundred years. What had been an antiquated process that rewarded court favorites and actively discouraged innovation by hungry, unknown inventors became, with the 1852 act of Parliament, an echo of the wildly successful American system, with its low fees and simple application process. Thus the Crystal Palace exhibition was a showcase not just for things, but also for the system that inspired inventors to get those things made and marketed. The American system.

If you were a dreamer, a world-summoning public event was the place to be. This was what you did with what you'd made, this was precisely how you took that first step down the road of making a fortune from it. The London event drew more than six million visitors to the eight-hundred-thousand-square-foot space that spread out beneath Sir James Parton's magnificent iron and glass edifice. Eight miles of exhibits—more than thirteen hundred displays—competed for the crowd's attention, yet many visitors couldn't tear themselves away from the demonstration provided by a brash, enterprising young American named Colt, last seen blowing up ships in harbors while people cheered and whistled.

This time, Colt took apart and then reassembled, over and over again, his repeating pistol. A repeating pistol! And one whose mechanics were so exquisitely simple and straightforward that you could deposit all of its parts in a box, mix them up in a confusing jumble and then put the whole thing back together again, part by part, in minutes. Which Colt confidently did, while the crowd, agog, leaned forward to watch. That communal awe was surely one of the reasons that Queen Victoria, surveying the rows and rows and rows of breathtaking exhibits on opening day, proclaimed it "the greatest day in our history." Not to be outdone, New York City held its own international exhibition a year later, on the site of what is today the New York Public Library.

In 1867, the International Exposition in Paris was a grand showcase for Samuel Morse's telegraph, as well as for the seventy-five other inventors who were trying to improve it with their own devices. Even though the telegraph was no longer new, it still seemed magical to the more than six million people who beheld it at the exposition. Imagine—a wire that could carry information from one place to another, just like that. The contraption seemed a bit clumsy at that point, and it was slow and rather unreliable—but *think* of it. A wire. A simple wire. And words suddenly were freed from having to be uttered face to face or slapped on a page and carried under your arm in a bundle of bound leaves. The words henceforth could fly through the air. *They could fly through the air.* Astonishment was the only plausible reaction.

At the International Centennial Exposition of 1876, held in Philadelphia for a six-month stretch, more than nine million people took in exotic marvels such as the sight of six thousand silkworms from China, busily doing their work, Daniel Webster's hand-built plow, and a revolutionary new floor covering called linoleum. Among the exhibits was an impressive array of Gatling guns; you can imagine small children daring each other to touch the sleek, bundled barrels, only to jump back with a frightened squeal. Overwhelming all other exhibits, though, was a coal-fired steam engine that sprawled in magnificent profusion in the middle of Machinery Hall, the main exhibition space. "An athlete of steel and iron" is how the Corliss engine looked to William Dean Howells. To underscore its importance, the engine revved to life during opening-day ceremonies on May 10, 1876, with a determined lever-pull by none other than President Ulysses S. Grant. He was joined on the podium by the emperor of Brazil, who happily lent a hand with starting up the huge machine.

First a choir sent aloft the stirring notes of the Hallelujah Chorus to rouse and inspire the crowd of more than one hundred thousand people, then Grant and the emperor did the honors, and the engine's two giant pistons shuddered to life. Fourteen acres of shafts, connected to the rest of the machines in the hall, surged with activity as power from the lumbering central engine throbbed through the massive room. The moment was magical; it made Americans busting-out proud of their country, and the lusty cheers created another sort of Hallelujah Chorus, this one aimed at a different kind of deity: the machine.

Grant's speech that day—uttered in the smelly, chaotic midst of vendors yelling "Ice lemonade!" and "Roast potatoes, roast potatoes hot!"—was both a pat on the head and a kick in the rear: "Whilst proud of what we have done," the president said, "we regret that we have not done more." More, most assuredly, would be done.

In 1893, it was Chicago's turn. The Columbian Exposition unfurled across the lakefront in all of its alabaster glory, brandishing to the world the Ferris wheel and Cracker Jacks and moving sidewalks and picture postcards and a bright paradise of electrical lighting—ten times more electrical lighting than had been on display at yet another Paris exhibition four years before. One out of every four Americans made their way to the Chicago fair to get a load of the more than sixty-five thousand exhibits spreading across some sixty-three million square feet. And like the London event in 1851, the Columbian Exposition wasn't just a matter of products and gadgets, of bright lights and sparkly objects. It was here that Frederick Jackson Turner, a history professor at the University of Wisconsin, presented a lecture titled "The Significance of the Frontier in American History" to colleagues who had gathered at the fair. It would become one of the most influential essays ever written about the nation's character and self-image. And it would demonstrate that fairs and expositions were not just gaudy spectacles, empty of intellectual content. The best-known and most widely anthologized chapter of Henry Adams's autobiography, *The Dynamo and the Virgin*, was written in response to his visit to the 1900 Paris exhibition, a visit that produced a constant series of double-takes for the sensitive Adams. In the "great gallery of machines," he wrote, he "began to feel the forty-foot dynamo as a moral force, much as the early Christians felt the Cross." Both awestruck and unsettled by the inventive genius on display, Adams realized that man "had translated himself into a new universe which had no common scale of measurement with the old." Fairs also boosted legitimate scientific inquiry; a year after the World's Columbian Exposition in Chicago, the city's Field Museum helped sponsor a two-year expedition to study transportation systems in more than twenty countries.

Sprinkled in between those globally famous blockbusters with their milling throngs of hundreds of thousands of people—from 1851 to 1889, there were more than fifty large-scale exhibitions—were the smaller but still thrilling public events: Robert Fulton's steamboat launch in 1807. The opening of the Erie Canal in 1825, during which colorfully decorated steamboats took their own sweet time heading to Albany, passing banks along which crowds of people gathered to wave and cheer, and cannon salutes thundered from each lock. The Boston Railroad Jubilee in 1851. The completion of the first transcontinental railroad in 1869. The first cable car trip in San Francisco in 1873. The opening of the elevated railway in New York City in 1876. The first roller coaster at Coney Island in 1884.

They occurred amidst music and ribbons and applause and a sense of manic excitement that could cut through a crowd like a runaway horse, quickening hearts and stoking imaginations and keeping everyone on their toes. It was the glorious opposite of the everyday. All was imbued with the delicious sense that anything—*anything!*—could happen, so you'd better show up and you'd better get there early and you'd better bring the children, too, and once you got there, you'd all better stay alert. Because this was the kind of era it was turning out to be: a time of quick change and wild discovery and well-organized flamboyance.

Even the grisly accidents and tragic snafus at such events were exciting—so exciting that they failed to put a damper on the grand tradition of public demonstrations. In 1844, during a test firing of an impressive new cannon known as the Peacemaker on board the warship U.S.S *Princeton,* a backfire resulted in the deaths of the secretary of state and the secretary of the navy. And at the Paris World Exhibition in 1855, Krupp, the family-run German munitions firm, reaped a publicity bonanza when its hundred-ton ingot of cast steel, ready for cannon making, crashed through the main showroom floor. It was a perfect metaphor for a bombastic and muscular new age: Nothing could contain the marvels to come. These were—quite literally—earth-shattering times.

Fairs, exhibits, celebrations, demonstrations. Indoors, outdoors. Small towns and big cities. All seasons. Every kind of weather. Drawing young and old, rich and poor. Go-getters and wastrels. Dreamers and cynics. A great many of the century's most important technological breakthroughs first made themselves known to the public at mass events, where the air crackled with the near-hysteria of collective anticipation. (Sometimes the high spirits had a bit of a nudge; Colt served up free brandy to those who visited his booth at the Crystal Palace Exhibition, and Fritz Krupp offered free champagne to potential customers during demonstrations of his massive weapons.) Many people got their initial glimpse of electric lights at a fair such as the 1883 Southern States Exposition in Louisville, where the Edison Electric Lighting Company had installed 4,600 lightbulbs.

Four years before, Edison had unveiled his creation—the incandescent lightbulb—at a public demonstration that was meticulously planned to tickle the press as well as the public. Sticklers later would note that Edison had not exactly "created" the lightbulb; rather he had "improved existing models, developing a filament which would glow consistently." But nobody pitched it as Edison was doing. Inventors had learned that it wasn't enough merely to dream up a new device; they had to push it, too. They had to woo their intended targets. They had to sell, sell, *sell.* Otherwise they might end up like Elias Howe, who had invented the sewing machine and patented it in 1846 but didn't make the kind of fortune from it that was made, in canny turn, by Isaac Merritt Singer.

It was Singer who sent sales agents scurrying across the country in the 1860s, visiting fairs and setting up special exhibitions to show women the wonders of a home sewing machine. The invention was one thing, but promoting the invention was another thing entirely. A Connecticut schoolteacher named Frank Holland devised an ingenious new fountain pen in 1879, but there is a very good reason why almost no one has ever heard of a "Holland pen" but a great many people know about a "Waterman pen": L. E. Waterman improved, manufactured, and marketed it. Creating a public appetite for one's handiwork was as important as creating the device itself. It was a lesson that Richard Gatling had learned many years before, as a young man in St. Louis sporting a newly patented seed planter and facing a circle of skeptical farmers.

Not every demonstration was for an entirely reputable product, of course. It wasn't all lightbulbs and locomotives. The nineteenth century was also the great age of the lively and pervasive fraud, when publicly touted patent medicine remedies such as Dr. Morse's Indian Root Pills and Bateman's Pectoral Drops and Turlington's Balsam of Life and Lee's Bilious Pills and Kline's Tooth Ache Drops and Judson's Worm Tea and Dr. Chilton's Fever and Ache Pills and Dr. Larzetti's Procreative Elixir were sold as surefire cures for what ailed you. It was the age of the notorious con and the silky swindle. An overheated sales pitch by a debonair quack could easily sway a batch of wide-eyed townsfolk. Before he settled down to gunmaking, Samuel Colt traveled from town to town as "the celebrated Dr. Coult [sic] of New York, London and Calcutta," administering nitrous oxide—enticingly referred to on his posters as "the Exhilarating Gas"—to gullible audience members. For a mere fifty cents, people could witness "innumerable fantastic feats" performed by Dr. Coult's subjects, once the "singular compound" had been administered. "Some individuals were disposed to laugh, sing and dance," Coult's posters noted, while others were prone to "recitation and exclamation," yet most displayed "an irresistible propensity to muscular exertion such as wrestling, boxing." And then there was the infamous Dr. Rutherford J. Worster, who operated out of Washington, D.C., during the Lincoln administration, cheerful purveyor of "Electro-Magnetic Insulators, Rejuvenating Elixir, Cough Syrups, Stomachic-Vegetable Electric."

It was technology, though, that most intrigued at these public demonstrations, it was *things,* it was the mighty engine and all the newfound sources of power. Different ways of getting things done. Different ways of harnessing the world's forces. Paramount was a radiant sense of possibility, a conviction that human ingenuity was ready to assert itself in a series of fantastic products and discoveries. The natural world with its mighty rivers and vast oceans could now be more than matched by the achievements of American builders and inventors, as a poet named Walt Whitman insisted in 1871: "I realize . . . that not nature alone is great in her fields of freedom and the open air, in her storms, the shows of night and day, the mountains, forests, seas—but in the artificial, the work of man too is equally great . . . in these ingenuities, streets, goods, houses, ships—these hurrying, feverish, electric crowds of men, their complicated business genius." The world was passionately caught up in what historian David Nye calls "the technological sublime," in the notion that what was once the exclusive province of nature, of mountain ranges and waterfalls, now could be replicated by human will and intelligence. A steamboat could be as lyrical and inspiring as a sunset. A locomotive, as splendid as a tree-fringed lake.

And a gun, of all things, could be beautiful.

It is the most effective implement of warfare ever invented.

There. A simple declarative sentence, unadorned, stripped of pauses and equivocation, scraped clean of hedging qualifiers. The first line of an 1863 sales brochure for the Gatling gun is a gem of unfussed, straight-ahead bravado. The rest of the broadsheet—unfurling beneath a pen-and-ink drawing of a Gatling gun, and the heading "Gatling Gun Or Battery," followed by the subheading "This Improved Fire-Arm is the invention of R.J. Gatling, of Indianapolis, Ind."—is arranged in strongly worded paragraphs of unembarrassed self-praise:

> *It can be discharged at the unprecedented rate of from 150 to 200 times per minute.*
>
> *The gun is simple in its construction; strong and durable, and it is not at all liable to get out of order from use.*
>
> *It is light and easily transported and would be found invaluable as a light field battery to be used in cavalry service, in the place of mounted Howitzers, or as a support to field artillery.*

Undreamed of, at this point, is the use of a machine gun as an infantry weapon, as something an individual might carry; that monumental operative and psychological leap will not come for decades. Straight out of the gate, it is regarded strictly as artillery.

> *It is confidently believed that no body of troops could be made to withstand the fire of such a death-dealing weapon, for the reason that men will not fight on such terms of inequality, or when there is no chance of victory.*

The aural appeal of alliteration—"death-dealing"—is tied to war's lethal logic: Who fights on when the odds seem overwhelming? Thus the element of intimidation is linked, early on, to the mere sight of a Gatling gun, to the fact that those gathered barrels seem to bristle with menace.

And then the advertisement zooms in on the particulars of the age: the Civil War. Gatling knows his market. He knows that a good salesman persuades his customers of the urgency of the purchase. It is not enough to point out why the customer needs it; rather, the customer must know why he needs it *now*.

The brochure continues:

> *The great object to be attained, and which every patriot should have at heart, is to crush the rebellion, and to do so, with the greatest possible saving of life and treasure.*

Next it asks and answers its own question, a staple technique of the salesman's and the politician's trade:

> *How can this be done? Only in one way—by crushing the military power of the rebellious States—and the sure means of accomplishing that result is to strengthen our armies; and the way to do that, is to arm our soldiers with this gun, and other improved breech-loading fire-arms.*

There follows an enumeration of what the brochure writer calls the "advantages" of the Gatling gun, rattled off with the confidence of a prosecutor making an airtight case:

> *1st. Its use will more than double the strength of our armies now in the field.*
>
> *2nd. It will lessen the number of men required, thereby removing many of the difficulties attending the procuring of men by draft, &c.*
>
> *3rd. It will save lives, wounds and sickness, by lessening the number subjected to the perils of war.*
>
> *4th. It will save immediately in treasure, by lessening the amount to be paid for bounties and pensions, arming, clothing, feeding and transporting the troops.*
>
> *5th. It will lessen the number of hospitals and hospital expenses—in a word, it will be a means of decreasing all the expenses of the war.*
>
> *6th. Its use will give to our armies greater mobility, enabling them to be more easily and more rapidly transported, handled and managed in the field.*
>
> *7th. It will give our troops an immense superiority and advantage over the rebels, because they cannot provide their troops with the same weapon.*
>
> *8th. If the government should adopt this fire-arm and the necessary metallic cartridge, and introduce it into the service, it will not be in the power of the rebels, in case they should capture the same, to use them against us, as it is impossible for them to manufacture, or procure, the cartridges in any way, and would, therefore, render the guns utterly useless in their hands.*

The brochure's big finish is an admixture of accusation, challenge, pugnacity, threat, and rhetorical questions, all served up with almost biblical-sounding cadences:

There is a fearful responsibility resting upon those who have the direction of this matter. The fact is, the Government has not yet availed itself of all the advantages, which a kind Providence has placed within its reach to crush the rebellion. It is, certainly, its duty to do so.

To this very hour we stand face-to-face with an enemy with arms very little, if any, inferior to ours! Is it not time for a change?

Across the bottom, the advertisement features the signed endorsements of the likes of generals and, of course, "Gov. O. P. Morton of Indiana," along with "many other officers now in the U.S. service."

Morton was no pushover. Even with Gatling's high-powered connections, even with the dreadful necessity of a war that seemed to go on and on like a bad dream, the Indiana governor had wanted to make sure. He hadn't relied solely on the splash Gatling's gun made in front of the townsfolk and the assembled military brass. Before Morton signed his name to any brochures, before the sales and marketing efforts could get off the ground, Morton had obtained a professional assessment of the new weapon. On July 14, 1862, several months after the public demonstration, the governor hired three firearms experts—a T. A. Morris, A. Ballweg, and D. G. Rose—to test the weapon for accuracy and durability. Their report was a rave: "The discharge can be made with all desirable accuracy as rapidly as 150 times a minute, and may be continued for hours without danger, as we think, from overheating."

Three and a half months after that private test, on November 4, 1862, Gatling received U.S. Patent No. 36,836 for his gun. The Gatling gun, its inventor boasted, "bears the same relation to other firearms that McCormack's reaper does to the sickle, or the sewing machine to the common needle." Now all he had to do was make the world believe it, too.

• • •

Later that year, in a decision that would come to haunt him for a long time, Gatling contracted to have the first six Gatling guns built in Cincinnati. It was a city he knew well from his days as a traveling man. Gatling had gone up and down the Ohio and Mississippi Rivers on crowded steamboats in the 1840s and 1850s, carrying advertising brochures for his agricultural implements, the drills and the hemp brakes.

The mighty thrust of steam travel had opened up river cities such as Cincinnati to flourishing commercial trade. From a marshy river bottomland thronged with an unruly tangle of white maple, dogwood, buckeye, tulip, pawpaw, sycamore, and gum trees flinging their branches out over the Ohio River, Cincinnati had risen, in less than fifty years, to become the business hub of the Ohio Valley and much of the Midwest. It was called the Queen City, and no wonder. Beyond the famed limestone quay that distinguished Cincinnati's crowded riverfront was an impressive-looking line of factories and foundries, standing shoulder to shoulder. Rising behind it was a succession of bright green hills. Urban and rural life still mingled in Cincinnati, sometimes to comic effect: Despite all that thriving and up-to-date industry, hogs ran wild in the streets, prompting the city's second and less majestic nickname: Porkopolis.

As amusing as the image of runaway pigs might be, however, Cincinnati made the most of its copious livestock. The city pioneered large-scale slaughterhouses decades before the Chicago stockyards became a factor in national commerce. By the late 1830s, Cincinnati had developed an efficient assembly-line system for the slaughtering of pigs. Working strictly by hand, without a machine in sight, a twenty-man team—each man assigned to a different part of the operation—could kill and prepare three hogs per minute. Each worker had "a special duty. . . . One cleaned out the ears; one put off the bristles and hairs, while others scraped the animal more carefully." The industrial miracle known as the assembly line worked with pigs and, a bit later in the same city, it worked with guns, just as it would work many years later in different cities with cars and computer chips. Cincinnati may have been a rough-hewn frontier town, where the porcine squeal was a kind of local anthem, but it was also a bold laboratory of industrial innovation. "All that there is of good or bad in American society is to be found there," Alexis de Tocqueville wrote of an 1831 visit to Cincinnati, "[E]verything there is in violent contrast, exaggerated; nothing has fallen into its final place."

Gatling's business partner was Miles Greenwood, one of Cincinnati's most famous citizens. In 1832 Greenwood had founded Eagle Iron Works, among the largest and busiest foundries in the western United States. During the Civil War, the barrel-chested, ferociously pro-Union Greenwood was a reliable source of government supplies; with only a day's notice, his shop manufactured a dozen anchors for pontoon bridges. His machines rifled the barrels of three thousand smooth-bore muskets desperately needed by Northern forces. At his factory, 150 field guns were cast out of bronze. Name your impossible task, set the bar improbably high, and Miles Greenwood could get it done. Miles Greenwood was your man.

Portraits from the time show a stern, balding, prideful-looking fellow with high cheekbones and the kind of facial hair that seems to have a mind of its own, growing wherever and howsoever much it chooses. Greenwood was very much a man of the era, entirely characteristic of the sort of individual produced by its churn and its vigor, a symbol of something bigger than just himself, a harbinger, a portent. He kept one eye on his bustling shop and another eye on the horizon. He was a builder and a visionary but a down-to-earth businessman, too, a man who could make things with his bare hands but who was always trying to figure out a way of getting bare hands out of the equation. Machine tools were the key, he knew. Making things one by one, in the old-fashioned, labor-intensive way, would never do for long. Not in this new country, the one that was spreading along the rivers and pushing its way west. Greenwood had been born in New Jersey, but his family moved west—didn't everyone head west, wasn't that the only direction that really mattered?—when he was a child, and by the time he reached his early twenties, Greenwood was an important man in this river town, a man with a solid and enviable reputation, a man who could make anything out of iron. Just you name it.

Once his business was off the ground, Greenwood turned his attention to public matters, serving on the Cincinnati City Council, expressing his opinion on civic issues large and small. People were inclined to listen. As the city grew, its newly hatched homes and stores and factories shooting up right next to each other like siblings sharing an attic bedroom, bumping and jostling, fire was a grave and constant threat. An early ordinance, in fact, had required all residents to keep a leather bucket handy in their dwellings, ready to be used for hauling water if a fire got the upper hand. Greenwood, paying attention to the latest technology, had a better idea. In 1852 he got together with two fellow Cincinnati businessmen—Abel Shawk, a locksmith, and Alexander Bonner Latta, who knew his way around a machine shop—and the trio built what is widely credited with being the nation's first steam-powered fire engine, a clanking, hissing, ten-thousand-pound behemoth that testified, once again, to the noisy triumph of human ingenuity and problem-solving pugnacity.

And it emanated, as did so many other technological developments of the time, from a combination of selflessness and self-interest: Yes, Greenwood was a public-spirited sort, but there was also the fact that his establishment had caught fire on several occasions. So the new fire engine, which could get a good stream of water going in less than ten minutes, was a savvy business move as well as a civic boost. For the people of Cincinnati, seeing it in action must have been like an instant holiday. "It had a square fire-box, like that of a locomotive boiler, with a furnace open at the top, upon which was placed the chimney," ran an admiring contemporary account. "The upper part of the furnace was occupied by a continuous coil of tubes opening into the steam-chamber above, while the lower end was carried through the fire-box, and connected with a force-pump, by which the water was to be forced continually through the tubes throughout the entire coil. When the fire was commenced the tubes were empty, but when they became sufficiently heated, the force-pump was worked by hand and water was forced into them, generating steam, which was almost instantly produced from the contact of the water with the hot pipes. . . . It is said to have played 210 feet through a thousand feet of hose, getting its supply from a cistern and afterwards when taken to New York on exhibition in 1859 it threw 375 gallons a minute, playing about 237 feet through a nozzle measuring an inch and a quarter." A gang of upstarts from the West, showing those New Yorkers a better way to fight fires. The country really was changing, its borders shifting, the interior opening up. The lid was off the box. Nothing could hold it back now.

Except, perhaps, for this Civil War, this protracted bloodbath that was threatening to undo all the prosperity and stability the young nation had achieved in less than a hundred years. Gatling and Greenwood both were staunchly on the Union side. That was among the many things they had in common. They liked to solve mechanical problems. They were self-made men. They had carved their fortunes out of the American wilderness with their own hands, not relying on family connections or fancy education. They were men of their time, but there was a timeless quality about them, too, a sense that you could set them down in any era and they would make their way forward. Hence Greenwood was the ideal partner for Gatling, or so it seemed; he appeared to be the perfect man to help the inventor turn out the first batch of working machine guns that the world had ever known—and just in time, of course, to be employed in the Northern war effort.

They were dreamers, skygazers, but they were businessmen, too. Merchants. Capitalists. Like Greenwood, Gatling never minded making a profit from his convictions. That was a sentiment shared by many notable men of the nineteenth century, including entrepreneurs such as J. P. Morgan. They were shrewd, restless men, men who never stopped inventing. (In Morgan's case, the inventions were generally new ways to make money.) Gatling and Greenwood looked at the world not as a fixed inevitability but as something fluid and supple, as a thing responsive to specific forces brought to bear upon it. The world wasn't rock; it was river. Gatling had switched from designing farm implements to making guns. And so in 1853, with the new steam fire engine ready for service, Greenwood had set up the first professional fire department in the United States, a corps of well-drilled men who were paid an annual salary of sixty dollars, with extra pay accruing to certain specialties: pipemen, drivers, engineers. And it all happened in Cincinnati. Not New York. Not Philadelphia. Not Boston. Gatling knew his man.

What he didn't know, however, was the disaster that loomed from the confluence of geography and politics. Cincinnati, Ohio, was right across the river from the state of Kentucky. No state was more contested in the early days of the Civil War than Kentucky, where Northern diehards and Southern hotheads commingled in vicious, simmering disequilibrium. At Cincinnati, runaway slaves often crossed the Ohio River from slavery-friendly Kentucky to the abolitionist stronghold of Ohio. Cincinnati was a fatal pivot point. Its position was so inherently dramatic, its politics so fevered, with the opposing sides in that furious, monumental dispute forced to coexist within constant and galling sight of each other, that it fired the imagination of young Harriet Beecher Stowe, who lived in

Cincinnati with her family in the 1830s and 1840s. The result was *Uncle Tom's Cabin,* the novel that rocked a nation and, many surmised, helped initiate the Civil War. Cincinnati was the mid-nineteenth-century world writ small: progressive and forward-looking but also agitated, raggedy-edged, ready to explode.

And it was there, in the city's volatile confines, that Gatling chose to have the fledgling models of his new gun made in November and December of 1862. It was a mistake, as he was shortly to learn. Gatling had underestimated the ruthlessness and ardor of the Confederate cause. Making the first Gatling guns in Cincinnati was a decision that would cost him in ways that he couldn't foresee, ways that would echo long after the last shout from the final fleeing rebel band at war's end. Yet Gatling had his reasons.

One night in December, just after Greenwood's men had finished assembling a gleaming row of six Gatling guns, priced at $1500 each, and gone home, flames shot out from the windows of the Greenwood foundry. Even the steam-driven fire engines couldn't save the factory or the Gatling guns. Greenwood had been warned repeatedly by Southern sympathizers from just across the river to stop making big guns for the Union troops. He ignored them. So this fire, unlike the earlier blazes that had plagued Greenwood's establishment, may not have been the result of crowded conditions and flammable construction materials. Nothing was ever proved definitively, but few had any doubt: Rebels had set the fire, leaving the Gatling guns molten heaps of ruin.

The same thing would happen to another gun merchant two years later, at the Colt Armory in Hartford, which also was turning out weapons crucial to the North: a mysterious and utterly devastating predawn fire. No readily apparent cause. Whispers that rebel saboteurs had finally made good on their threats.

For Gatling, who had financed the making of the newfangled guns out of his own pocket, it was a catastrophe. Another kind of man might have quit right then and there, having lost money and time and momentum. Gatling kept going. He took his business to the Cincinnati Type Foundry. Thirteen Gatling guns were made there in 1863. Two years later, after the war, he would move again to the Cooper Fire Arms Manufacturing Company in Philadelphia; in 1865 and 1866, this was the factory that turned out Gatling guns. In 1866, he'd make a shift to the Colt factory, which became the permanent home of the Gatling gun.

Fires, dislocations, lost dollars, soured business relationships. If Gatling was discouraged, if he ever wrung his hands in frustration, you never saw it. He hid his emotions well. He had mastered the entrepreneur's unflappable veneer, the coolly radiant confidence. Some anxious feelings occasionally leaked out in a letter or two, when Gatling urged a sales representative to take a stronger hand after a negotiation had fallen through, or inquired after an unpaid invoice. Mostly, though, he had the nineteenth-century business-man's constant cordiality, that knack of enduring setbacks without flinching or faltering. Whatever it took, and wherever it took him, he would get his guns made. And then he would get them sold.

If you had walked into the Cincinnati Type Foundry in early 1863 you would have found, fixed securely atop their two-wheeled field carriage mounts, a baker's dozen of gleaming new six-barreled Gatling guns. These models, unlike the original six destroyed in the fire at Greenwood's facility, were designed to utilize the new .58 caliber metallic bullets instead of paper cartridges, a vast improvement.

Gatling got his guns made, the war went on, but it was not to be. He was too late with his innovation. His gun was too easily confused, perhaps, with a similar-looking weapon by another inventor, a weapon unrelated to the Gatling gun, that also purported to shoot multiple times but never panned out. Still, Gatling pressed hard for the government to add the Gatling gun to its arsenal. So did an enthusiastic O. P. Morton. Yet all that passionate support, all that letter writing and exhibition sponsoring and test-firing and testimonial issuing, made scant headway. It was a seed flung on flinty ground. The Union brass was dead set against anything but traditional armaments. A long war was making everyone nervous, risk-averse, much more conservative than they already were. The army's chief of ordnance, Lieutenant Colonel James W. Ripley, was notoriously contemptuous of newfangled weapons. He was weary of the promise-the-moon brochures that cluttered his in-box. War was serious business; he had no time for backyard tinkerers and their crazy claims. No patience for their clumsy-looking contrivances.

He was exasperated, Ripley wrote in a stern 1861 missive, at "the vast variety of new inventions, each having, of course, its advocates, insisting upon the superiority of his favorite arm over all others and urging its adoption by the Government" and added that "this evil can only be solved by positively refusing to answer any requisitions for or propositions to sell new and untried arms."

Lincoln, though, was intrigued by new technology. He was a bit of a backyard tinkerer himself, having invented a contraption to hoist riverboats up and over shoal water. As president he had argued on behalf of early attempts at creating multiple-firing artillery—not Gatling guns, which were still on Richard Gatling's drawing board at the time—but when the guns proved troublesome and unreliable, even dangerous to their operators, in actual battle, Lincoln backed off. He had, Lord knows, a few other things on his plate. Henceforth he'd leave the gun buying to the experts: Ripley and his staff at the ordnance department. And they were fed up with nontraditional firearms. That frank prejudice helped them ignore the sterling reports that came in from official trials of the Gatling gun, trials conducted by trusted military men. Two test firings at the Washington Navy Yard in 1863, on May 20 and July 17, produced this brief, blunt encomium from the supervising officer: "Mechanical construction is very simple, the workmanship is well executed, and we are of the opinion that it is not liable to get out of working order." That sounds like faint praise, but given the dismal record of other new weaponry in the heat of actual Civil War battles, it was a rave. It was distinctly promising. The gun worked "admirably," according to the ordnance officer who had put it through its paces at the trials. But no one in authority was listening.

The U.S. Army did finally adopt the Gatling gun for official use and ordered one hundred—but not until August 21, 1866, after the war. If the Gatling gun worked so well, why weren't the men in charge impressed enough to change their minds when it would have mattered? The time of its first introduction was, after all, a desperate one in the war; as a distraught Lincoln noted, after listing the calamities that had befallen his troops by 1862, "The bottom is out of the tub." If Gatling's innovation truly was odds-changing, world-altering, why didn't the Union Army, so beleaguered at the time Gatling offered it, embrace the weapon with grateful gusto? The answer lies in the folds and creases and crevices of history, where bits of personalities, scraps of circumstance, and flecks of coincidence are stashed, waiting to be plucked out and assembled into a more complete picture. "It didn't work right" is the charge often hurled against the Gatling gun, a charge that is not even remotely accurate, but the accusation has lingered for over a century and a half now. What gives it its peculiar staying power? And yet it is indisputably true that during the Civil War, the Gatling gun was a weapon that nobody much wanted.

CHAPTER TWO

A World of Mornings

As it is with the individual, so it is with the nation.

—Theodore Roosevelt

There he comes. He's riding due south through the thicket of trees and confused-looking underbrush, startling birds and small animals and keeping a practiced lookout for bears, which are known to lurk in these parts. Colors are stark and particular. Everything is green and high and wild, and crowded with mysteries underfoot. The sky is as blue as an Easter bonnet, a light and powdery shade that tails away to white at the edges. The heat sticks to his neck like a poultice.

It is 1833. Richard Jordan Gatling is fifteen years old, solid of build, with a round face, a shy and hopeful smile, small slitlike eyes, and lank dark hair that he splits with a left-side part. He's on his way from his family's cotton plantation—a big and successful one, with some twelve hundred acres and some two dozen slaves—in northeastern North Carolina to the closest town, Murfreesboro, five miles away. He's starting work there for his great uncle, a clerk in the common pleas court, his first job away from the farm. He knows that just ahead runs the Meherrin River, which he and his horse will cross by ferry.

He has a quick mind and ample hands. He likes to think about things, to figure out how physical forces are generated and where they go and why. A toothed wheel fascinates him. Gravity and gears and friction and inclined planes: What could be better? What could be more exciting? He, like his country, is feeling the pull of the beautiful unknown, is falling under the spell of the new.

He resembles his country in other ways, too. He's ambitious. He's hungry for an unnamed something. He's all arms and legs and sprawling, unfocused want. Winding his way through the heavy foliage that shoulders right to the edge of the narrow dirt road, a road that's really just the hushed rumor of a road, he passes through patches of sunlight and shadow and sunlight again. The unruly rummage of the trees overhead, which allows the light only sporadic access, guarantees it: sun, shade, sun, sun, shade, sun. As he moves through the days of his future he will, in a similar way, merge with his country's destiny and then part from it, not a symbol so much as a whole and unique human being, and then he will blur into abstraction again, so closely does his biography interlock with America's. Thus will he merge and part, merge and part. He's burning with energy, truly comfortable only when in motion, only when leaning irredeemably forward.

He is a young man in a young country. Between 1815 and 1839, the average age in the United States is sixteen. And in roughly the same batch of years, the country will undergo changes of amazing and unsettling magnitude: some good, some bad, all relentless and swift. This is the time of "the establishment of the ideological, social, financial and technical basis of an industrial economy in the United States," notes historian Richard Slotkin. The same energetic trajectory will sweep up Gatling and carry him aloft and along: He was born on his family's farm in an isolated area of the South in 1818, and by 1844 he will be on his way to the thriving and cosmopolitan city of St. Louis.

First, though, he'll work for a while in his great uncle's Murfreesboro office, the object of this day's journey along the dusty road from farm to city. He's a careful and conscientious employee, but his mind is elsewhere. He has, you might say, an elsewhere kind of mind.

It's an ideal age for an original thinker. The perfect atmosphere for the kind of person who pants and squirms with impatience at the old order of things. America in the early 1800s: It's a world in gorgeous flux. A world of mornings. Nothing is certain. Everything is up for grabs. It's an age when water isn't water anymore. Rather, it's water—but it isn't *only* that. It is more than itself. It is also, and astonishingly, power.

Like the alchemists of old—but on a surer footing, certainly, on a foundation of rigid fact and scrupulous testing instead of whimsy and superstition—some people in the first few decades of the nineteenth century are beginning to understand that things can throw off peculiar and fascinating shadows of themselves into the future. Familiar entities are shifting into useful new forms. Light becomes color. Wire becomes words. Wind, music. And water, power.

Gatling knows about water. He was born by the north bank of the Meherrin River, and he spends a good portion of his boyhood sounding its depths—and his own as well—with experiments and observations and also time spent just plain fooling around on homemade boats, but it's a practical sort of fooling around. A forward-looking way of fooling around. And it's a fate he can't escape: His family is mechanically minded. His kinfolk are all practical problem solvers, through and through. They're inventors, right down to their callused fingertips.

Richard Gatling's father, a prosperous farmer named Jordan Gatling, obtains two patents in 1835; one's for a device that thins cotton, the other is for a rotary cultivator. Richard Gatling's brother, James Henry Gatling, will receive a patent in 1871 for a process that converts fallen pines into usable timber. James Henry Gatling also will invent a primitive airplane before anybody has ever heard of a couple of brothers named Wright. It's a clan of tinkerers. Of constantly preoccupied dabblers. Of shade-tree mechanics and armchair eccentrics, of men with pencils angled behind their ears, men who always seem to be just seconds away from reaching for a piece of paper and sketching up plans for new ways to perform ordinary tasks: lifting, pushing, pulling, hoisting, separating, hauling, moving, and, unashamedly at the end of it all, making money.

The adolescent Richard Gatling, likewise, is fond of designing things, of refining and improving farm implements and tools and new machines. Devices that do all kinds of work better and faster and cheaper. After a year of working for his great uncle, he'll spend another year teaching school nearby, a common occupation for young men of the day from well-regarded families, even though they lack a college education. Young Daniel Webster does the same thing in his New Hampshire neighborhood.

But there is something twitching inside Richard Gatling. Call it an itch or an urge or an instinct. He wants to see more of the world than he has, more than what's been set on the plate in front of him by birth and geographical accident, more than just the rivers and farms and dull daily familiarity of Hertford County, North Carolina.

Hertford County. It acquired that designation in 1759, as homage to the Marquis of Hertford. Early on, explorers from the original Jamestown settlement dipped down into this area on several occasions, nosing around, scouting out another good place to start up a community. Colonel Hardy Murfree, who made his name in the Battle of Stoney Point, a 1779 Revolutionary War skirmish in New York, obligingly lent it to the city of Murfreesboro. Two better-known Murfreesboros—one in Tennessee, one in Arkansas—came after. Murfreesboro, North Carolina is the original. Among its distinctions, along with being the birthplace of the inventor of the machine gun, is the fact that Dr. Walter Reed lived there as a child, from 1855 to 1857. The physician who went on to discover that yellow fever is transmitted by mosquito bite, saving millions of lives, and to have his name bestowed upon the famous army hospital near Washington, D.C., lived in Murfreesboro when his father, a Methodist minister, was assigned to a church in town. Walter Reed later returned to marry his childhood sweetheart from Murfreesboro, Emilie Lawrence.

There was a time, in the late 1700s and early 1800s, when Hertford County was an important inland port. A tangle of small rivers persistently interrupted this land, as if the Atlantic Ocean, realizing it had reached firmament, still didn't quite have the heart to stop, and certainly not all at once, and so it thinned and stretched and curved and probed as far as it could, only gradually letting itself be funneled into smaller and smaller threads, from rivers to creeks, finally giving up, with a sigh, in a series of marshy swamps. From the Atlantic, ships could enter the Ocracoke Inlet or the Albemarle or Pamlico Sounds, and then travel up the Chowan River, and thereby reach the busy wharf of Murfreesboro. If their captains were so inclined, the vessels could keep on going, skimming across tributaries such as the Chinquapin and the Potecasi.

Once the sailing vessels yield to their faster, noisier successors, Murfreesboro and vicinity do see their share of steamboats. As the century heats up in the 1830s and 1840s, however, as everything gets bigger, from ships to ambitions, the river can't keep pace. It is too narrow, too shallow. The world begins to pass Murfreesboro by.

Left behind are a number of small farms and a distinctive way of life, one that most emphatically—and to the men in charge, most unapologetically—includes slavery. In August, 1831, when Richard

Gatling is thirteen years old, a slave on a plantation a few dozen miles north of the Gatling farm, just over the Virginia border, leads a brief, bloody revolt. Nat Turner's act shocks and terrifies the neighborhood. The Gatlings and other local families move into Murfreesboro, huddling there for some two months until Turner and his comrades are caught and hanged.

Routine is suspended. The familiar rhythms of Southern life are shattered by the violence. It is a short but profound foretaste of what lies ahead for the nation as a whole: the terrible whirlwind that is waiting over the next hill. There will be a reckoning. The country cannot go forward without one. Deep in the woods of North Carolina and Virginia, down on the plantations where, as a contemporary novelist envisioned it, "scraggly pine groves stretched across the landscape . . . [and] a smoky haze hung over the land, and crows cried dismally from afar," the Gatlings and their neighbors feel the vibration of approaching hoofbeats. It will take another three decades from this moment in 1831 for the full fury to strike, for the South to be called to account for the ghastliness of slavery, but when it gets there, it will be sweeping and catastrophic. At that point, there will be no escape into a nearby town. No waiting it out. They have had their hint. They know what's coming.

Four years later, in the fall of 1835, Richard Gatling is seventeen years old and restless. He obtains a license to operate a store at Frazier's Cross Roads, just down the road from the Gatling plantation. He does that for four years, in between the running of errands for his father. But the inventor in him can't be kept at arm's length for long. He comes up with a plan for a new kind of propeller for watercraft. The United States government has put out the call for just such a device. Naval officers want a propulsion system that can work under the waterline; otherwise, it's vulnerable to enemy fire.

Oars and paddle wheels, the way many ships are powered at the time, are cumbersome and inefficient.

Gatling gets the idea for his propeller, he will recall many years later in an 1891 interview with a North Carolina newspaper, during a trip to Norfolk. He's there conducting business for his father. Down by the waterfront, he sees a trial run of a new propeller design: two paddle wheels at the back of the boat, boxed in by a small compartment to protect them during their revolutions. It still doesn't look right to Gatling. Doesn't seem efficient. And it certainly lacks the elegance of the truly wondrous inventions, the kind that set off fiery pinwheels in his imagination.

He thinks: Why not attach a series of rotating blades, at oblique angles, to the axis of the turning shaft? Why not, that is, generate force from a screw-like device with fluted surfaces instead of the flat surfaces of a paddle wheel?

He rides home on horseback from Norfolk and gets busy. He builds a propeller with four wooden blades sprouting at forty-five-degree angles from a central shaft. It looks a bit like a windmill. Gatling tests it in a pond on the farm. It works. He's ready. Because of the kind of family he comes from, and because of the kind of country he lives in, Gatling knows his next step. He may be young, he may be a kid from the farm, a kid nobody ever heard of, but that doesn't matter. He knows exactly where he needs to go.

If you had a good idea in the early nineteenth century, and if you were fairly certain of its commercial prospects, you had but one destination: the Patent Office. It didn't matter if you possessed scant formal education. It didn't matter if you couldn't spell terribly well, or didn't have an illustrious family posing importantly in the background, or lacked fancy clothes or polished manners. If you could prove that what you'd dreamed up was useful, and that it was yours, you could make sure nobody else came along and stole it out from under you.

In the spring of 1836, an eighteen-year-old Richard Gatling travels to Washington, D.C., where he learns that he is too late—but only by a hairsbreadth. Someone else has come up with a similar invention and filed an application for a patent just a few months ahead of him. John Ericsson's screw propeller will revolutionize water transport, becoming one of the great Swedish inventor's most significant discoveries, a list that also includes steam boilers, caloric engines, and a process for making salt from brine. The principle of propulsion by screw propellers is still the main means by which marine vessels are powered today.

Gatling, no moody brooder, goes home and starts again. He dreams and he draws. He never stops. It's a lifelong habit of mind. In the next two and a half decades, before he invents the gun that will bestow upon his name a perverse immortality, Gatling's restless, creative brain comes up with a passel of ideas. He is, like the irrepressible Irene Lapham in the William Dean Howells novel *The Rise of Silas Lapham*, someone who's "wide-awake, every minute." There's a new variety of plow, a cotton cultivator, a washer to tighten gears more effectively, and that's just for starters. Even before his gun makes him famous, his other inventions make him rich.

They make him rich because they are effective and they answer specific needs, but there's more to it than that. They are invented in a country in the throes of a bright new idea: the patent. If there is to be a democracy governing political affairs, then similarly, there must be a democracy of the imagination, or so believed the upstart country's founders. And the headquarters for those democratic imaginings, at the time of Gatling's initial trip to Washington, D.C., in 1836, is located on the west side of Blodgett's Hotel on E Street NW, between Seventh and Eighth streets. It had moved to Blodgett's in 1810, after calling the Treasury Office home for the previous decade. The federal government had transferred its headquarters from Philadelphia to Washington in 1800.

In 1840, four years after Gatling's first disappointing visit with his small model of a new kind of propeller, the Patent Office moves into a splendid new headquarters. In the century that is just hitting its full stride, the Patent Office will become a top tourist destination in the nation's capital. Because inventors are required to submit models with their applications, the Patent Office is filled with gadgets. Many visitors will concur with a rapturous Walt Whitman, who dubs it "the noblest of Washington buildings," housing, another observer says, "the nation's curiosity shop."

In 1836, though, the Patent Office consists of just a few small rooms and a handful of clerks and a fiercely singular conviction. With his early trip there as a hopeful teenager, Gatling joins the great brigade of American inventors who are remaking the world, one idea at a time.

They are able to do this because of something called—with utterly misleading blandness and sterility, with no hint of its propulsive magic and explosive potential—the patent system.

If a country can be said to possess a soul, then America's is the patent system: the simple, fair method of staking claim to a new idea and getting the chance to make money from it. And like a soul, it is invisible—few people, when asked what makes America uniquely successful, would say, "the patent system"—and yet it is crucial to what the country becomes. When George Washington signs the patent bill into law on April 10, 1790, the United States is the first country in the history of the world to grant inventors the legal right to profit from their creations. A copyright law for written work is passed shortly thereafter, on May 31, 1790. One by one, throughout the nineteenth century, other nations of the world follow suit. When their leaders see what the patent system does for America, how magnificently it kindles creative fires and sparks economic growth, they end up adopting the same kind of patent protection—known as the American system—for their citizens.

Even before that 1790 law, patents matter. The Constitution, ratified in 1787, sees to that. Article 1, Section 8, declares: "Congress shall have the power . . . to promote the progress of science and useful arts by securing for limited times to authors and inventors the exclusive right to their respective writings and discoveries."

Simple words, but words of stunning import. Because the patent system is what pushes and inspires and elevates the early United States, quickly distinguishing it from Europe and the Old World. In the Old World, the game is rigged from the start. Everything depends on family name and power and influence, on pleasing kings and queens. Everything depends on ancient arrangements, on intricate networks of patronage and favoritism. In the Old World, you are what you were at your birth, and that is all you can ever hope to be. In America, however, you are what you dream of being. Here, an inventor's fate is limited not by family name or social standing, but by the boldness and ingenuity of the vision. It is a revolutionary notion, a furious new form of energy, and it hustles the country forward. Such frenetic and lucrative activity surely is exhausting to watch from afar, and even a tad exasperating. Looking back on the busy century of American progress, a British banker admits as much: "The United States," he says in 1899, "is rather too enterprising for the peace of mind of Europe."

Mark Twain, holder of three patents himself and a writer with a sly and unsurpassed sense of the American soul, has a character in *A Connecticut Yankee in King Arthur's Court* note that "a country without a patent office and good patent laws was just a crab and couldn't travel anywhere but sideways and backwards."

By making patents available to anyone with a good idea—regardless of birth station or bank account—George Washington and his fellow Founders create a new class of striver: The earnest amateur. The individual armed not with formal degrees or important connections, but with skills and ambition. Yet it is more than just the accessibility of patents that proves momentous in the new country. It is also the fact that patents bestow financial incentive, because the patent system backs up inventors' rights to profit from their creations, preventing others from stealing the idea and reaping the commercial benefits. This is key: The desire to make money, to get ahead, is recognized in the very earliest laws of the new land, a point that was not lost on a rising politician named Abraham Lincoln. "The Patent System," he declares in 1859, "added the fuel of interest to the fire of genius." In the same speech, Lincoln—himself the holder of a patent—sees patents as the crucial distinction between the United States and the nations that long predate its founding: "The great difference between Young America and Old Fogy is the result of Discoveries, Inventions, and Improvements." Even earlier, Thomas Jefferson recognizes the almost mystical role played by the new country's patent system: "The issue of patents for new discoveries," he writes, "has given a spring to invention beyond my conception."

Initially, Jefferson himself is charged with examining and ruling on patent applications. The patent law has created legal protection, but not a Patent Office; the latter will come about in 1836, when the number of patent applications just keeps rising, requiring a separate bureaucracy and a full-time staff of examiners. In the meantime, the applications are sent to a board made up of the secretary of state—Jefferson—the secretary of war, Henry Knox, and the attorney general, Edmund Randolph.

It can't be a task Jefferson wholeheartedly relishes. An inventor himself, he is famously conflicted about patents, inherently opposed to government-enforced monopolies. He fears they may end up stifling creativity and innovation. He takes out no patents on any of his inventions. Yet he acknowledges how essential patents are: "Certainly an inventor ought to be allowed a right to the benefit of his invention for some certain time," he writes. "Nobody wishes more than I do that ingenuity should receive liberal encouragement. . . . In the arts, and especially in the mechanical arts, many ingenious improvements are made in consequence of the patent right. . . ." For a fee that varies between $4 and $5, and by submitting a small model of the proposed invention, an American citizen—aliens won't have the right to patent their creations until 1800—can be granted a patent lasting fourteen years.

That relatively low fee is important. Elsewhere in the world, patent fees in the nineteenth century are roughly ten times what American inventors must pay. High fees are another way of keeping patents in the hands of the wealthy and powerful, of locking out the average citizen. In Great Britain, which only gradually comes around to the American system, extending a patent requires an act of Parliament—and that, in turn, requires entrenched political influence that the regular inventor does not possess. A British patent at this time is awarded "by the grace of the Crown"—an elegant-sounding phrase that really means that the government can favor the rich and well connected, no matter how worthy or innovative the new invention might be.

It's different in the fledgling United States. Even before the Constitution is ratified, inventors get a jump start. Several colonies hand out patents. Samuel Winslow receives a patent in 1641 from the state of Massachusetts for a salt-making method he's perfected. Five years later, Joseph Jenkes receives the emerging nation's first manufacturing patent for his design of a mill to turn out scythes.

Under the freshly minted 1790 patent law, the first American to receive a patent is Samuel Hopkins of Pittsfield, Vermont. It's for a new way of making potash, a form of potassium carbonate created by burning trees, used in the making of soap and other products. Four years later, an inventor named Eli Whitney patents a device known as the cotton gin. In 1834, Cyrus H. McCormick patents a machine for reaping wheat. Two years after that, Samuel Colt is granted a patent for his repeating pistol. Four years later, it's Samuel Morse's turn. He patents his telegraph. In another six years, Elias Howe patents his sewing machine. In the meantime, Richard Gatling has made his first trip to the Patent Office. He's rebuffed, but he'll be back. In 1857, the United States issues 2,920 patents, some 35 percent more than Great Britain; Great Britain, of course, has many more people than the United States at this time. In Russia, the patent total is 24. In Prussia, it's 48. Clearly, the American patent system is working. It is motivating inventors with the possibility of renown and fortune, spurring them on as nothing else in history has ever quite motivated the species before. The Founders had human psychology down cold. An era of unprecedented inventiveness has dawned, a gloriously turbulent age of new machines and fresh methodology, and the Richard Gatlings of the world will make the most of it.

Not that it's easy. Not that it comes without peril and sacrifice. Through the years the Patent Office itself faces a variety of calamities. As if to underscore the fact that the patent system is at the center of the American story, its architectural embodiment repeatedly finds itself in the middle of the nation's history in the nineteenth century: the War of 1812, the Civil War, Lincoln's second inaugural ball.

In the last week of August, 1814, British troops march toward the United States capital city with a simple goal: *Burn it. Burn it all to the ground.* At the Capitol building, marksmen fire into the row of windows on the eastern side to thwart any snipers that might be lurking there. Then the British soldiers batter down the doors. They loot the place, ending their pillage by ripping up the woodwork and using it to set fires. Rockets launched into the Hall of Representatives destroy the glass-tiled roof and ignite even fiercer flames. The congressional library lovingly compiled by Jefferson, 740 volumes strong, is turned into ash.

The next stop for the British troops is the Treasury building, then the U.S. arsenal, then a barracks. All are burned. Stunned observers in Baltimore, thirty-four miles away, can see the jumping flames, the rising smoke. Arriving at the president's house, the British troops soak rags in oil, set fire to them, and hurl them through the windows, but not before raiding the absent President Madison's wine cellar and helping themselves. The next day, the undefended State and War Departments lie in the British sights, appealingly vulnerable: more fires, more destruction. The mayor of Georgetown, meekly waving a flag of truce, begs the British soldiers not to burn his town. Admiral George Cockburn agrees and notes, with amused disdain, that he is in effect providing more protection than the citizens have received from Madison.

Finally, the invaders get to Blodgett's Hotel, home to the Patent Office. It is here that the British forces are stopped in their tracks, not by a cadre of menacing soldiers, not by an iron wall of fierce-looking artillery, but by a persuasive analogy.

William Thornton, first head of the Patent Office, steps forward to tell the British officers that if they destroy *this* building, they will be no better than the marauders who destroyed the library in Alexandria. He implores the soldiers not to burn "what would be useful to all mankind." Housed in the Patent Office are not only the patent applications themselves, but also the small models inventors submitted. The devices and inventions at the Patent Office—the contraptions that are the tangible manifestations of thoughts and dreams—belong not just to America; they belong to humanity. Wounding this structure wounds the future.

The argument works. The British back off. The moment reveals just how important the idea of inventiveness is to the nation's early development, how profoundly revered. In the aftermath of the British attack, as fire-gutted government buildings smolder in sullen ruin, the temporarily displaced members of Congress meet in the Patent Office.

It is a fitting contingency plan. The Patent Office has enormous symbolic power for Americans in the new century. The Patent Office—more than the Capitol, more than the president's residence, more than the headquarters for the various bureaucracies—is the repository of the country's hopes and ambitions. Here is where good ideas can be turned into great fortunes, where vision and persistence will pay off.

Little wonder, then, that the Patent Office becomes a notable tourist destination in the nineteenth century, at one point luring more than one hundred thousand visitors per year to gaze at the models that inventors deliver along with their applications. There are nutmeg graters and brick-making machines and dog-powered treadmills and bedbug traps and lard lamps and soap holders and boot stretchers and clothes pins and chimney caps and tricycles and doorbells, and there are also things such as steam pumps and sewing machines. Turn a corner, and you'll likely bump into something that you never could have envisioned—but that someone else could, and did, and wants to make and sell.

By the time Gatling actually receives his first patent—it's for an ingenious new kind of seed planter—on May 10, 1844, hopeful inventors no longer go to a makeshift office in the cramped wing of a ramshackle hotel on what was then the outskirts of the young capital. In

1840, a new Patent Office opens. At last, this palace of the imagination truly looks like a palace, too.

The new Patent Office at the corner of Seventh and F streets, across the street from the post office, is a massive white Greek revival edifice composed of Virginia freestone and Maryland marble. Pierre L'Enfant, the original designer of America's capital city, had reserved the space for a national church; in a way, that's just what the Patent Office is. It's a secular church that celebrates the bright and endless mysteries of the American imagination.

The new Patent Office takes up two full city blocks. It's not even close to fully finished in 1840, but it opens anyway, because the Patent Office needs the space. Viewing the impressive place just after its unveiling, a *London Telegraph* correspondent calls it "magnificent in proportion and design." Wide steps rise in stately grandeur toward great Doric columns. Through the door flows a steady ribbon of not only eye-on-the-horizon inventors ready to file patent applications—the number tops two dozen a week at this point—but also awestruck visitors, tourists killing time, curious citizens who wander around the display-filled cabinets in the model rooms, hands clasped behind their backs, a bit dazed, a bit overwhelmed, nodding appreciatively at this visible proof of what the human mind can do if sufficiently motivated.

The model room is the best show in town, many say. It's where you see the tangible manifestation of the country's future, nicely laid out for leisurely viewing. "All who love new & strange sights," a visitor to the Patent Office in 1844 enthuses in his diary, "should spend a day in Washington Patent Office." Here, he continues, one can find "improvements of every kind in agriculture & mechanics besides inventions for useful & ornamental purposes, steam engines, miniature railroad cars, waggons, coaches, saddles, hats, umbrellas, chairs, sofas, knives, pistols, houses, fences, things for all purpose and all forms. No room in the world can exhibit more evidences of thought than this. We are called an inventive people but none know how true is the saying until he has carefully examined the Patent Office. You feel rejoiced that such useful inventions are fostered by our government and that talent can generally meet its reward."

A rambling visitor to the Patent Office who's not in a hurry could take in "models of John Ericsson's screw propeller, Alfred Vail's printing telegraph, Jonas Chickering's grand piano with iron frame, Stephen Fitch's turret lathe, Elias Howe's sewing machine, or Richard Hoe's rotary printing press. Jumbled together with these were hopeful gadgets designed to expedite almost every conceivable object from the writing of letters to the propulsion of ships—there was even an 'electrifying machine' and a device [called] the 'sacrificator.'" Charles Dickens calls the Patent Office "an extraordinary example of American enterprise and ingenuity." Patents are stepping stones into the future, set at intervals corresponding with the great leaps forward enabled by the American inventive genius. Thomas Edison is granted his first patent in 1869, for a stock ticker that employs telegraphic technology; in the ensuing four decades, he will average two patents a month. Edison's mantra: "A minor invention every ten days and a big thing every six months or so."

After 1836, inventors' designs are submitted to a staff of patent examiners. The notion of employing a panel of professional reviewers catches on in other parts of the world, too. Among the most famous patent clerks ever to do the job is a scruffy young employee of the Swiss patent office named Albert Einstein. The experiences of Einstein, the twentieth century's greatest scientific mind, within a patent office are testimony to the crackling energy and optimism of such a place. A patent office is not just another bureaucracy. Here, the best ideas of the day come surging past, a great gush of innovation and aspiration, a parade of potential. Einstein calls the patent office "that worldly cloister where I hatched my most beautiful ideas." He is hired on June 16, 1902, and for the next seven years, Einstein sits on a stool and reviews patents, a task that, he later explains, enables him "to see the physical ramifications of theoretical concepts." These are Einstein's most creative years; this is the magical period during which he is remaking physics by tying energy to mass and coming up with his special theory of relativity.

Switzerland, like virtually every progressive country in the world, frankly imitates the American patent system, seeing how well it is working for the United States, how it stimulates economic development by creating incentives for inventors. Before 1888, the Swiss government had not offered patents. The products for which the Swiss were known—watches and chocolates—and the country's small population mean, its leaders believe, that a patent system won't make much difference. But by the late 1880s, Americans have been awarded more than two thousand patents on watches and watch making, and the Swiss realize they need to keep up, to push ahead. An increasingly interlocked global economy means that nations aren't isolated anymore. Competition comes from everywhere. So the Swiss government establishes a patent office, and one of its bright young employees turns his time there—as he reviews, day after day, the work of other scientists and inventors, as his mind then goes its own way—into nothing less than an upending of time itself, of traditional ways of looking at the universe. The patent office is Einstein's laboratory. And his launching pad.

In 1886, a Japanese envoy named Takahashi Korekiyo travels around the world, comparing patent systems. He studies the patent laws in European countries and in America. His conclusion: "We said, 'What is it that makes the United States such a great nation?' and we investigated and we found it was patents, and we will have patents." In 1888, Japan enacts its first patent law. An American-style staff of patent examiners is a central part of the legislation.

• • •

For a Richard Gatling, and for many of his countrymen, the patent system is the great spark, the push, the provocation; and the financial opportunities enabled by its protections are the lure and the beacon. Between 1844 and 1862, Gatling obtains nine patents for agricultural implements; his inventions include a hemp brake, a rotary plow, a lath-making machine, a gearing machine and a steam-driven marine ram. Gatling matches his devices to markets. He doesn't invent things just for the sake of inventing things. He invents things to make money. Practicality—the aim of earning a fortune by solving an immediate problem—is fused with poetry: the desire to create something wondrous, something new in the history of the world.

Devices dreamed up with patents in mind will transform the daily life of the nineteenth century, from private functions—the disposal of bodily wastes, the washing and ironing of clothes—to public ones, such as getting from place to place and communicating over long distances and fighting wars. The inventions to alter and streamline and simplify these activities come, for the most part, from the minds of amateurs. The patent system is the amateur's best friend, because in the United States, patents are issued simply on the worthiness and originality of the invention itself. Other factors are irrelevant. As Lincoln declares in an early speech, "All creation is a mine, and every man, a miner." Amateurs such as Gatling—people, that is, without college degrees or certification or training—thrive in a way that will not be possible in the next century, when codified credentials begin to matter inordinately. The era of the amateur doesn't last long: "With every decade" in the nineteenth century, says historian Paul Johnson, "the tyranny of professional qualifications and apprenticeships was becoming more entrenched." Yet while it does last, it is gloriously productive.

The word "amateur" sometimes is used derisively, pronounced in a dismissively pejorative tone, but it derives, fittingly, from the

Latin word meaning love. "These were the days," notes historian Robert Bruce, "when a lone inventor could succeed without capital or formal training, so long as he had imagination, mechanical ingenuity and a few tools. Invention was a pastime still wide open to the general public."

The rise of the amateur is a singular phenomenon in the world's history. "A collective fervor for invention seems to course through this period," observes technology historian Siegfried Giedion. "In the seventeenth century the inventive urge was possessed by a limited group of scholars—philosophers and savants like Pascal, Descartes, Leibniz, Huygens." But then came the American patent system, which spread the inventive spirit "over the broad masses. . . . Invention was in the normal course of things. Everyone invented, whoever owned an enterprise sought ways and means by which to make his goods more speedily, more perfectly, and often of improved beauty. . . . Never did the number of inventions per capita of the population exceed its proportion in the America of the [18]60s."

This sense of egalitarian hope and infinite possibility suffuses the national spirit, as amateurs in the nineteenth century come to be distinguished by their earnestness and their good intentions. They believe in self-improvement as well as the improvement of society; in fact, they believe that the improvement of individuals, one by one, is the surest route to the overall betterment of the nation. They form clubs and learned societies. They meet to read Dante, as a group of intellectuals in Cambridge, Massachusetts, does in the 1860s, at the behest of Henry Wadsworth Longfellow. Or they get together to "argue questions of theology and philosophy," as do members of the Nineteenth Century Club—a New York–based gathering of business tycoons, including Andrew Carengie, and leading intellectuals of the day—also beginning in the 1860s. They attend public addresses and trade opinions about profoundly universal topics. They read newspapers and magazines with energetic relish.

They organize amateur musical gatherings and dancing clubs. Parents who can afford the cost often hire tutors for their children in fields such as fencing and French and penmanship. Public libraries are regarded as essentials, not frills. The filigree of life—literature, music, manners—is not restricted to the upper classes; the middle class, too, desires its share of such civilizing influences. The lyceum movement, with its lectures and its energetically sincere devotion to intellectual inquiry, blooms and spreads. It is joined, in 1874, by the Chautauqua movement: more self-improvement, more belief in wisdom and community.

Beginning as a Methodist summer camp meeting in upstate New York, the Chautauqua ideal catches fire. Organizers start publishing home-study courses, correspondence courses, a newspaper. In the early 1880s, the *Chautauquan* hires a young writer named Ida Tarbell, whose deep belief in social justice—a distinct feature of the age, a conviction that if people are only informed of wrongdoing they of course will try to eradicate it—soon sets her in opposition to John D. Rockefeller. Tarbell's carefully reported articles on Rockefeller's monopolistic business practices become the standard for the emerging field of muckraking journalism. The spiritual root of that field, though, is pure nineteenth-century earnestness. The belief that people truly want to do the right thing. The confidence that perpetual self-improvement through knowledge is not only possible, but desirable. Authors, theologians, military figures, and notable thinkers—everyone from Herman Melville to P. T. Barnum to Clara Barton to Ralph Waldo Emerson to Elizabeth Cady Stanton to Frederick Douglass—are booked for lengthy tours in the United States. They generally draw large, appreciative crowds. People want to know what's going on in the world. They want to furnish their minds the same way they furnish their parlors: with practical utilitarian items, yes, but also with touches of beauty and poetry.

• • •

They want to make large fortunes, they want to succeed—oh, yes—but American entrepreneurs in the nineteenth century want more than that, too. They display "a shared esteem for high civilization and its masters," notes a historian of the period's business giants, and they enthusiastically support "programs of cultural uplift." Carnegie, in the midst of making millions in his steel mills, brings Herbert Spencer and Matthew Arnold to the United States for lecture tours to inspire the general public. If the amateur ideal finds its first manifestation in the sky-blackening factories of enterprising nineteenth-century Americans, it finds its second in the pristinely elegant parlors of the lyceum movement. Here, culture is regarded not as the exclusive birthright of the upper classes, but as the gift to all. From its very founding in 1870, New York's Metropolitan Museum of Art declares in its charter that its goal is to edify the masses, not just to provide swanky settings for masterpieces: The institution "is not surpassed as an educational power among the people by any university, college or seminary of learning." The museum offers regular classes in crafts and the mechanical arts to anyone who wants to attend.

The message is clear: In the United States, culture is accessible to all. It is a carefully packaged commodity that can be acquired through undertaking the right course of study, through joining the right clubs, through paying attention to the correct things. Many of the financial behemoths and inventive wizards of the age pursue refinement the same way they seek profits: steadily, methodically, with an unyielding belief in goal-oriented systems and solid progress. Good taste is displayed through objects, and objects can be bought. The Hartford home of Samuel Colt is stuffed with expensive and exotic art objects from all around the world, with sculpture, stoneware, portraits, and gilt mirrors, reflecting the wealthy arms maker's desire to "light the way to an idealized future," historian William

Hosley notes. In the late 1850s, "with day-to-day operations at the armory increasingly delegated to others, Colt turned his attention to . . . creating a home that validated the law of progress." He and his wife, Elizabeth, indulged "a passion for art, a yearning for self-improvement." Art, Colt believes, can "mitigate the unsettling effects of industrialization," Hosley writes. The gunmaker understands that concepts such as the lyceum are "instruments of the cosmopolitanism and worldliness that, if applied to the raw talent of American genius could not fail to multiply national power and prestige."

The lyceum. The word itself has an earnest, emphatic ring to it, a rhetorical radiance, a sense of purpose and positive thinking. It's a word that instantly conjures up an image of honest striving after abstract knowledge–knowledge sought not only to ensure wealth or notoriety, although there's nothing wrong with wanting to get rich, but also to enable growth as a good person. "Lyceum" was coined from a Latin word that in turn harkens back to the Greek designation for the sun god. Like the very knowledge it disseminated, the word itself spread and took on new associations throughout nineteenth-century America. It started out in the 1820s referring to a slate of public lectures, then began to encompass as well the buildings erected specifically to hold the lectures, and finally came to mean the concept itself, the whole forward-leaning, perspective-expanding, soul-enriching panoply of it all. Josiah Holbrook, an amateur scientist and publisher in early nineteenth-century America who believed that everyone ought to learn as much as they could about everything it was possible to know, helped spread the lyceum concept across the country, following the great wave of population growth from East to West. By 1831, Cincinnati had a lyceum; the next year, Cleveland and St. Louis joined, and in 1834, the citizens of Chicago could attend a lyceum. When Springfield, Illinois, began one in 1838, the organizers chose as speaker a gangling local attorney named Abraham Lincoln. His topic: "The Perpetuation of Our

Political Institutions." Austin, Texas, established a lyceum in 1841, followed by Galveston, Houston, and Brownsville in 1845, 1848, and 1849, respectively. By the 1850s, lyceums could be found in cities in Kansas, Nebraska, and California.

You can use the growth of the lyceum movement to follow the westward march of the great themes of the nineteenth century, the relentless appetite for big spaces and epic adventures. While this march often is seen as America's enthusiastic embrace of the idea of manifest destiny, historian Frederick Merk argued that it is not that at all. It is not the all-encompassing greed for new land that motivates America's westward push, but idealism, or "mission," as he called it. It is the radiant, selfless conviction that, as American diplomat Albert Gallatin puts it in a report in the 1840s, "Your mission is to improve the state of the world, to be the 'model republic,' to show that men are capable of governing themselves, and that the simple and natural form of government is that also which confers most happiness on all, is productive of the greatest development of the intellectual faculties, above all, that which is attended with the highest standard of private and political virtue and morality." It is, in short, the idea of the lyceum.

A country that lacks the venerable educational traditions of the Old World—the ancient universities, the great libraries, the museums, the soaring cathedrals—still recognizes the importance of knowledge. Here, though, everyone shares in the knowledge. It isn't for an elite few. In America, the age of the amateur means that many people gladly pay admission fees on a regular basis to hear lecturers speak on subjects about which they think they ought to know more. Their lives are works in progress, no matter what they have already accomplished. As Holbrook notes in the weekly magazine he publishes in the 1820s, *Family Lyceum,* the word "expert" is not a term of snooty exclusion; expertise lies within reach of all. "It seems to me," Holbrook writes in an 1826 article, "that if associations for mutual instruction in the sciences and other branches of useful knowledge could once be started in our villages, and upon a general plan, they would increase with great rapidity, and do more for the general diffusion of knowledge, and for raising the moral and intellectual taste of our countrymen, than any other expedient which can possibly be devised." It's a new idea in the world: a club that opens the door to anyone who knocks.

"Front seats: a few old folks,—shiny-headed,—slant up best ear towards the speaker,—drop off asleep after a while, when the air begins to get a little narcotic with carbonic acid," is how Oliver Wendell Holmes describes his experience as a lyceum speaker. "Bright women's faces, young and middle-aged, a little behind these, but toward the front,—(pick out the best, and lecture mainly to that.) Here and there, a countenance, sharp and scholarlike, and a dozen pretty female ones sprinkled about." There is, to be sure, a trace of condescension in this passage, a hint of the weary intellectual spreading his riches before the bovine mass. The old folks, after all, "drop off asleep after a while." But there is also a suggestion of the hungry eagerness with which many people come to the lyceum. Those "sharp and scholarlike" faces are aimed at the speaker with hope and expectation. Through the lyceum movement could be traced "the measured footstep of an advancing civilization," according to prominent lyceum lecturer Thomas Wentworth Higginson in an 1868 manifesto. What's at stake is the very identity of the country, which in large measure is being created, night after night, lecture by lecture: "It is an exciting life," Higginson continues, "thus to find one's self moving to and fro, a living shuttle, to weave together this new web of national civilization."

And it is not just audience members who benefit from the addresses. In the lecture tour of 1895 and 1896, the one that ultimately takes Mark Twain from Cleveland to London, with stops in between in places such as Crookston, Minnesota, Great Falls, Montana, and New Whatcom, Washington, Twain finds joy: "Lecturing is gymnastics, chest-expander, medicine, mind-healer, blues-destroyer, all in one." There's something exhilarating about making people laugh, and making people think, especially when you can make them do both at once, which is the essence of Twain's playful genius, and also the essence of the lyceum ideal: entertainment with a serious, self-improving purpose.

In both lecturers and audiences, what counts, what stands out, is the earnestness. Always, the earnestness. It shines forth from the lyceum movement. The earnestness is a little touching, perhaps even a little childish-seeming, in its bottomless belief that the world can be transformed one questing mind at a time, that history is something you can get at—and thereby manipulate, setting it going the right way round—through information and good intentions. The earnestness is embedded in the proudly jingoistic title of a brand new magazine that begins publication August 28, 1845: the *Scientific American,* the nation's first mass-circulation magazine, and one that will prove instrumental in Richard Gatling's career and in the careers of other inventors and entrepreneurs. In its original incarnation, the *Scientific American* is a one-page broadsheet with the latest news from the Patent Office. As the number of subscribers increases, so does the page count; by the 1860s, the magazine includes handsome drawings of the innovative devices born of American ingenuity. The January 12, 1867, issue features a picture of the Model 1866 Gatling gun, displaying Gatling's most recent design change: an arched front frame, to reduce the chances of an errant strike from an exiting bullet. Throughout the last half of the nineteenth century, portraits of and stories about the Gatling gun will show up over and over again in the *Scientific American.* And the magazine is to have a sad link, many years down the road, to the very last day of Richard Gatling's life.

The *Scientific American* is born just ahead of a surge of other new magazines with national circulations, as Americans prove just how ravenous they are to keep up with things, how united in their desire to improve themselves. Between 1825 and 1850, the number of magazines published in the United States jumps from less than one hundred to some six hundred. A year after the *Scientific American* prints its first issue, the Smithsonian Institution opens, led by that ultimate amateur, the self-effacing Joseph Henry, the man who perfects the use of the electromagnet for long-distance telegraphy. The questions earnestly debated at the lyceum and the Chautauqua gatherings and in the pages of the new mass-circulation magazines drive right to the heart of the amateur ideal: What does it mean to be a person of integrity in the modern world? How can one reconcile individual success and achievement with the fate of the masses?

Judging from their letters and from the recollections of their contemporaries, Gatling and his fellow inventors and businessmen in nineteenth-century America don't believe that getting rich can be the whole goal. More is at stake than private fortune. And from this spirit will arise the world's first working machine gun—developed to make a profit, most certainly, but profit cannot be the exclusive end. His gun also is developed, Gatling will note, to shorten wars and save lives. The lyceum ideal informs the making of the Gatling gun, just as it does the creation of so many products and processes in the nineteenth century. Machines will liberate humankind. Machines will make us better, more efficient; they will free us for nobler pursuits.

Such a conviction may sound foolish and naïve to twenty-first-century sensibilities, but Gatling and his ilk really did seem to believe it, and this belief—poignant, even ridiculous, in the cold retrospective light of twentieth-century wars and strife—is their motivation, the driving piston of their desires, just as much as are dollars and power and palaces.

• • •

On November 8, 1877, a group of extraordinarily wealthy businessmen meets in New York to laud J. P. Morgan's father, Junius, for his lifetime achievements. After a lavish, champagne-laced dinner of oysters, foie gras in pastry shell, partridges with truffles, and baby hens, among other heaped delicacies, New York Governor Samuel J. Tilden scoots back his chair, stands up and toasts the financier for his having exemplified the highest possible ideal: a "consciousness that human society is better because we have existed." This is the purest strain of nineteenth-century thinking: Wealth and notoriety are fine, but the true measure of worthiness lies in doing good for others. As famed preacher Henry Ward Beecher proclaims in 1860, "Everybody knows that wealth, rightly employed, is as the right hand of God."

To be sure, many will discount the sincerity of these pious sentiments; Morgan and his ilk, some say, only pretend to care about such high-minded ideals, while amassing their dizzyingly immense fortunes. Yet we don't need to discern the secret hearts of the great nineteenth-century magnates—unknowable, in any case—in order to get a measure of the age. The fact that they even purport to care about the public welfare, that they consider it imperative to proclaim a belief in noblesse oblige, is enough. There is, in the atmosphere of the nineteenth century, a growing conviction that simply to become the most successful country in the history of the world is not enough. More is required. A nation blessed with unusual good fortune is a nation obliged to use that good fortune for another kind of good: good deeds. Otherwise, why was this special nation allowed to prosper so? Surveying America's constant interrogation of its own motives, historian Niall Ferguson notes that the United States "believes that heaven intended it to free the world, not rule it." But how to know when power is used justly, and when not?

These questions are the first stirrings of a debate that haunts the country down to the present day, when discomfort with America's military might—the word "empire" often is energetically refuted, despite its obvious reality—sometimes prohibits a fair assessment of the contributions of armaments innovators such as Richard Gatling. In the lyceum, the public ruminations over the American character provide an early taste of the creative challenges to come: How to reconcile manifest destiny with the indigenous people who are displaced in the wake of it. How to be a world power bent on imposing its will, without denying other nations the right to self-determination. Ferguson, noting its origin in the late nineteenth century, calls this "the paradox that was to be a characteristic feature of American foreign policy for a century: the paradox of dictating democracy, of enforcing freedom, of extorting emancipation." The United States, Ferguson and others have noted, is an "empire in denial." A rich, powerful country that—thanks to the ideals burnished in the nineteenth century—finds its riches fastened inextricably to a conscience, its power bound to an uncomfortable ambivalence about wielding it in the world.

Clubs, groups, learned societies: They didn't originate in the United States, of course. They simply find a new expression in a new country, a land where the doors swing open a bit wider, where the air does not seem to be quite so twice breathed. In late-eighteenth-century England, a group of men that includes Erasmus Darwin, Josiah Wedgwood, James Watt, and Matthew Boulton gather on a regular basis to discuss the latest scientific theories of the day. Their explorations, their inquiries, will set important parts of the foundation for the Industrial Revolution. This group is "not made up of aristocrats or statesmen or scholars but of provincial manufacturers, professional men and gifted amateurs," a historian notes. They call themselves the Lunar Club. Lacking specific technical training in fields that intrigue them, they possess something even more essential: a passion for knowledge.

In countries with rigid class structures, such as Great Britain, the means to act upon that passion is restricted to relatively small segments of the population, a fortunate few. In the United States, though, where the patent system encourages inventors from all backgrounds to pursue scientific and technical interests, the results are scintillating. What begins in the drawing rooms of a smattering of eccentric men in late-eighteenth-century England takes hold much more forcefully, and produces important results, in the nineteenth-century United States, where the patent system works its practical magic.

As historian Paul Johnson has noted, the Industrial Revolution does not emanate from the serene and leafy quadrangles of Oxford or Cambridge, or, for that matter, Harvard or Yale. It is the work of amateurs. It is not the professors or the professionals who change the nineteenth-century world, but the curious and the determined. The untutored dreamers who back up their dreams with hard work and ingenuity. It is a world in which a painter named Samuel Morse, who considers himself at best a "moonlighting inventor," devises—along with the telegraph—a piston pump for fire engines and a machine to cut marble. A world in which a self-taught physicist named Joseph Henry can not only become an eminent professor at Princeton University and discover the key to make Morse's telegraph work, but also serve as the first director of the Smithsonian and science adviser to President Lincoln. A world in which a Peter Cooper—with a flimsy if not downright nonexistent education and not a technical credential in sight—can invent a washing machine, a mechanical lawn mower, and the first iron locomotive, known as Tom Thumb.

• • •

It is a world in which a Robert Fulton, who starts out his life penniless and without prospects, can claim ownership of the idea for a steamboat and then nab nine other patents to boot, including one for an armored ship and another for an underwater cannon. A world in which a young man from Maine named Chester Greenwood, armed only with a good idea—why not strap a piece of wool across the ears, to protect them from the harsh winds of those infernal Maine winters?—patents it and then establishes a factory that in the 1870s turns out some four hundred thousand pairs of earmuffs annually, and then obtains more than a hundred other patents, for such things as a steel rake, a shock absorber, and a self-priming spark plug, all without any technical or business education.

A world in which James C. C. Holenshade, a confirmed autodidact, is able to make a killing in the hardware business throughout the Midwest, then take all of his knowledge and drive, all of the expertise about mechanical things that he's spent his life accumulating and plow it into promoting a weapon known as the Cincinnati Breech-Loading Cannon, an early hint of the great changes coming in armaments. A world in which J. N. Reynolds, explorer, lecturer, editor, a "natural amateur, a renaissance man with what seemed to be illimitable interests," holds the public so spellbound with his tales of adventure—*Volcanoes! Cannibals! Mighty whales!*—that he persuades the nation to support expeditions just for the sake of discovery, not only for the chance of financial and scientific booty.

This is also the last great age of the generalist, before the rise of specialization that will come to characterize the twentieth and twenty-first centuries. This is the now-vanished world in which a grand sweep of knowledge is deemed acceptable, before narrow specialization becomes the norm. This is the time when "science and art were not separated," writes historian Jenny Uglow of the late eighteenth and early nineteenth centuries. "You could be an inventor and designer, an experimenter and a poet, a dreamer and an entrepreneur all at once without anyone raising an eyebrow." Men such as the great Western explorer John Charles Frémont are not only adventurers but also geologists, astronomers, botanists, mapmakers, engineers. They are interested in everything. Like Reynolds, they cultivate an assortment of passions and many different skills. Moving restlessly in the very air of the nineteenth century is the sense that the world is discoverable, that its phenomena can be measured and understood, that men—and it is mostly men at this point—have a responsibility to unmask its secrets, to harness its great power, to profit from its marvels.

Profit: That is the watchword. By the midpoint of the nineteenth century, the Patent Office is the headquarters for the American dream, the central clearinghouse for the nation's creative energy and drive. The majestic white rectangle is, by now, a popular thoroughfare thronged by the nation's brightest minds and most ambitious personalities. "Though the patent office's available space had trebled since the Mexican War, it had not kept up with Yankee fecundity in mechanical invention," a historian writes. "The number of patented mechanical inventions began to rise in an exponential curve which did not pass its peak until the first quarter of the twentieth century."

The Patent Office is a portal, but it is engulfed by a larger portal: the Civil War. Everything that happens in nineteenth-century America must pass through the transformative fire of the Civil War. And here, too, the Patent Office plays a role. Until regular quarters can be found, 1,200 volunteers from Rhode Island are housed in the ornate halls before being sent into the fray. In an even grimmer link, at one point the corridors of the Patent Office are jammed with wounded and dying Union soldiers. From October 1861 to January

1863, the Patent Office is used as a hospital; the torn bodies are oddly juxtaposed with the tall, glass-fronted cabinets stocked with wonders. As the war worsens, military officials in Washington are scrambling for places to put the casualties. The stricken men are delivered, on stretchers wet with blood, to churches, hotels, colleges. Arriving by torchlight, usually, it seems, in heavy rain, they are conveyed to any location where floor space is available.

Walt Whitman, who has visited his brother George on the front lines, who has seen the wounds this war inflicts, wants to help. Poets aren't doctors or nurses, but there is much he can do. In his notes for a Civil War chronicle never published in his lifetime, Whitman leaves a vivid portrait of the scene at the Patent Office when it is pressed into service as an emergency hospital: "The vast area of the second story of . . . the Patent Office was crowded close with rows of the sick, badly wounded, and dying soldiers. They were placed in the three very large apartments. I went there several times. It was a strange, solemn, and—with all of its features of suffering and death—a sort of fascinating sight. . . . Two of the immense apartments are filled with high and ponderous glass cases crowded with models in miniature of every kind of utensil, machine, or invention it ever entered into the mind of man to conceive, and with curiosities and foreign presents. Between these cases were later openings perhaps eight feet wide and quite deep, and in these were placed many of the sick; besides, a great long double row of them up and down through the middle of the hall. Many of them were very bad cases, wounds, and amputations. . . .

"It was indeed a curious scene at night when lit up," Whitman continues. "The glass cases, the beds, the sick, the gallery above, and the marble pavement under foot; the suffering and the fortitude to bear it in the various degrees; occasionally from some the groan that could not be repressed; sometimes a poor fellow dying, with emaciated face and glassy eyes, the nurse by his side, the doctor also there, but no friend, no relative—such were the sights but lately in the Patent Office."

The war comes to the Patent Office, but the Patent Office also goes to the war. A Patent Office clerk named Clara Barton leaves her desk and rushes to the bloodiest sites of the conflict, becoming "unofficial ministering angel to the Army of the Potomac's wounded—and conscience to the Army's medical department." She will be instrumental in the founding of the Red Cross. On September 17, 1862, during the terrible carnage at Antietam—the casualty total for both sides will top 17,000—Barton arrives in her ramshackle wagon filled with medical supplies. So close is she to the front lines that she is almost hit by a bullet while nursing a wounded soldier.

On March 6, 1865, the ball for Lincoln's second inauguration is held at the Patent Office. Instead of dying soldiers, the marble floors are dotted with dancing couples in fancy dress, who pay $10 each to be there. Proceeds go to the war effort.

No death groans now. Only music from two bands and a sense of relief among the four thousand guests who are there to toast Lincoln's reelection, an event that had seemed, at one point, distinctly unlikely, given the war's grim duration. Lincoln and his wife arrive just after 10 P.M. They pick at the buffet supper and stay until 1:30 A.M., although many guests remain until 4. "I could not help thinking," muses Whitman, who stops by, "what a different scene they presented to my view a while since, filled with a crowded mass of the worst wounded from the war, brought in from second Bull Run, from Antietam, and from Fredericksburg. Tonight, beautiful women, perfumes, the violins' sweetness, the polka and the waltz—then the amputation, the blue face, the groan, the glassy eye of the dying, the clotted rag, the odor of wounds and blood, and many a mother's son, passing away unattended there." In the Patent Office in 1863, the massed casualties had seemed like an odd counterpoint to the magnificent array of inventions; now, just months later, there is a similar dissonance between the memory of the dying men and the waltzing couples. The Patent Office embraces it all. It is a spectrum of the nation's most hopeful and most harrowing moments, its fair pastel morning and its bloodred sunset.

On May 10, 1844, it is still very much morning for Richard Gatling. The twenty-six-year-old has just been granted his first patent: No. 3,581 for a seed planter. Like all of the machines he will invent, it is simple but clever. Walking behind the device, which looks a bit like a wheelbarrow with a sharp protruding blade along the bottom that digs a furrow for the descending seed, a farmer can plant his crops in uniform rows. This is a great improvement over the broadcasting method, in which the farmer flings the seed in all directions.

Gatling is on his way. He will receive forty-three patents in his lifetime, for contraptions such as a dry-cleaning machine and a steam-driven tractor and an improved flush toilet. His most famous invention, of course, the one for which he is known, is aimed not at improving life, not at enhancing it, but at destroying it, and that is the way it is with a democracy of the imagination, with the patent system itself: It takes all comers. It responds to human initiative, sinister or wholesome. Like that other kind of democracy, the kind that sometimes results in third-rate leaders and short-sighted laws, the kind embraced by a country that proclaims the primacy of human rights and yet, for far too long, condoned slavery, the patent system brings all sorts of things to fruition. It enables all manner of milestones, the good and the bad, the necessary and the superfluous, the marvelous and the monstrous, too. This mix, this tension of opposites, gives nineteenth-century America its distinctive flavor. And its distinctive tragedy.

Nothing seems tragic at this moment, however. The wide world beckons. Gatling is leaving home, mounting his horse for the daunting and perilous 737-mile journey from Murfreesboro, North Carolina, to St. Louis, Missouri, a place he's heard about, a new market, a part of the country he's never seen before but surely wants to. He's excited. It is a young man's sort of adventure. There's no dirt to scrape off his boots from earlier journeys of any magnitude. His imagination is busy with all the things he intends to create or perfect next: plows and seed drills, mostly. Hemp brakes, gearing machines. Big, practical things. Things of obvious utility. It's what he knows best. On the patent he receives for his seed planter, Gatling lists his address as Murfreesboro, North Carolina. It won't stay that way much longer.

If you ask him, Gatling will tell you that agricultural implements are the objects destined to carry his name forward. He'll use what he knows: wheels, gravity, revolving cylinders. And yet he senses, too, that he has a capacity for more, a depth yet to be sounded. Inside him, the faint shape of an utterly new invention—one that has nothing whatsoever to do with crop yields or sustainable land use, but instead involves death and domination—bides its time, waiting for the irrefutable necessity of its creation. As described by a contemporary novelist writing about 1870s America, this particular kind of concavity is a "sensitivity to the unknown that makes it a . . . specific unknown . . . as if we discern in the darkness only the dim risen quality that draws us toward it." A dark shape is rising in Richard Gatling's mind. He rides out to meet it.

CHAPTER THREE

Land of the Second Chance

Any man of push and energy could make his fortune or go emphatically broke.

—Howard Mumford Jones

The universe is sensitive to the merest touch and therefore it is possible to set wheels in motion that shall outrun the world.

—John Wanamaker

He headed west, and no wonder. If you wanted to go, West was where you went. "The future lies that way to me, and the earth seems more unexhausted and richer on that side," insisted Henry David Thoreau in 1862. "Eastward I go only by force; but westward I go free."

And what a world it was, the one you passed through on your way there in the early nineteenth century. Lush forests, clear rivers. Infrequent towns. Half-hearted roads. Few signposts. Just hunches and momentum. Just a whisper: *Onward.*

It was the great undiscovered country, equal parts geography and mythology, and it was the land through which Richard Gatling passed on horseback as he made his way to St. Louis. The cities of the East Coast—blushingly young by European standards—seemed old and arthritic when measured against the limber and frisky

West, the West of constant if high-risk opportunity, the West of go-right-now bravado, which was often and accurately equated with foolhardiness.

West meant west of wherever you were standing at the moment. In the first half of the nineteenth century, West meant Ohio and Indiana as readily as it meant California and Oregon. (Illinoisan Abraham Lincoln was routinely referred to as a Western politician.) Spreading out like the four corners of a broad green blanket with pegs placed on the Atlantic coastline and—moving clockwise here—down to Georgia and then across to the Great Plains and up to Canada was a single magnificent forest, thousands of miles of nobody-knew-what. It was, notes historian Barbara Freese, "one of the largest stretches of woodland ever to grow on the planet." It was, many geologists believe, the oldest forest in the world, and certainly the most diverse. America's rain forest, some have dubbed it. It was "the continent's seedbed, its mother lode."

If you squinted and concentrated in those days, you could see for almost ninety miles. By the late twentieth century, sulfur dioxide emissions from the burning of fossil fuels would reduce that view to less than fifteen miles on an average day. But throughout Richard Gatling's young manhood, it was ninety. No wonder so many people packed up and lit out for the West; they could literally see what they were missing.

"Always Americans looked westward," wrote historian Bruce Catton. "The nation was built on hope, ambition, and a contradictory bundle of dreams, and the pattern seldom varied. . . . The men who first crossed the ocean and the men who followed or went beyond them in later years had their eyes on a spot real or imagined, but always somewhere over the western horizon. . . . By the middle of the eighteenth century the restless vanguard which always drew the nation westward in its wake was crossing the wooded Appalachians into the vast, rich basin of the Mississippi. Once this region took shape in men's imagination as the new American West, the country's future was assured."

It would all change, of course, and relatively quickly, and the untrammeled West would be altered beyond recognition in mere decades. The speed of the change meant that a protective national nostalgia would be packed tight around the memories of those early days, the dry ice of enchanted narrative, of fantastic yarns, of tall tales. Larry McMurtry has pointed out that what we think of as the West—the West of legend and anecdote—vanished in a single generation. Jessie Frémont, wife of the explorer John Charles Frémont, summarized the velocity of the loss in a poignant letter to her husband: "All your campfires have become cities."

More than a century and a half later, we can only guess at what life must have felt like for Gatling on his journey west in 1844, during that brief instant in the world's biography when the last massive swath of undeveloped forest was up for grabs: the green munificence ranging across Pennsylvania and Ohio and Illinois and Indiana and nudging its way into Missouri. Leaving the thin corridor of the East to head west at that time was a journey from a specific and predictable something to a fantastic could-be-anything. Southern Indiana was a "frontier wilderness, heavily forested and sparsely settled . . . a land of dense somber forest, teeming with wild game, broken only by the log cabins in isolated clearings or by meandering wagon trails that linked the cabins to straggling village communities." As late as 1850, portions of Illinois had only two to six people per square mile.

Gatling's new home would be St. Louis, the last outpost before the "silence and the unknown" of the rest of the continent, as Jesse Frémont phrased it in another letter to her husband. Gatling was leaving behind his family, his friends, a settled life of routine and predictability. Which, to some, sounds richly desirable, while striking others as a mild kind of hell.

He didn't leave right away. He didn't depart the moment he was old enough to formulate his own intentions and declare them outright. He had tried storekeeping, just down the road from his hometown. He'd given it a fair shot. He'd tried to fit in. For a fee of six dollars, Gatling received a merchant's license, signed by the sheriff of Hertford County, North Carolina, on September 2, 1840, which granted him the legal right to sell "goods, wares and merchandize [sic]" for one year. He took notice of local affairs, so that he could discuss them with customers. He cast his first presidential vote, for William Henry Harrison, in 1841. But storekeeping didn't suit him. A dusty counter at a county crossroads in the rural South. Humid afternoons, stretching on forever and ever. Sleepy contentment.

No, no, no. His head was too full of all the things he wanted to build. Things he thought he could sell, thereby building a great fortune. He was a restless young man. He was brimming with energy and purpose and hope. He had caught his country's peculiar fever. There was no cure.

Gatling left home during what would come to be known as the "Roaring Forties," a nod to the tumult and chaos of the age. In the summer of 1845, a year after Gatling reached St. Louis, the term "manifest destiny" was coined by John Lewis O'Sullivan in the pages of the *United States Magazine and Democratic Review,* the kind of periodical that functioned like a pitchfork in the backside. "Manifest destiny" summarized a mind-set, a perspective. An upward chin thrust of an attitude. In 1846, after three decades of peace, the United States was mixing it up again; this time, it was the Mexican War. After years of quiet stasis, the nation was on the move.

But it had to move more slowly than it would've liked to, for the most down-to-earth of reasons. You couldn't get anywhere without considerable effort and dismal experience. The roads were wretched. In the early 1840s, the railroad was still too new, its scope too limited, to be a regular factor in travel. If you took the roads, you went on horseback or by stagecoach. If you took the rivers, you went by flatboat or by steamboat. All ways were difficult and uncomfortable and time-consuming. Stagecoaches might have seemed to be the safest, because other people were all around, but that wasn't necessarily true. Side curtains rarely kept out snow, wind, and rain. Soft tops to the coaches made accidents perilous. One hapless man traveling from New York to Cincinnati in 1829 reported that his stagecoach had overturned no less than nine times. The sorry state of transportation was more than an inconvenience; to some, it represented a stain on the American character itself. How could such a progressive country tolerate such god-awful roads? "To the New England mind," Henry Adams recalled ruefully, "roads, schools, clothes, and a clean face were connected as part of the law of order or divine system. Bad roads meant bad morals."

Today, a trip from New Orleans to New York can be accomplished in a flight of a few hours. In 1846, when an actor named Joseph Jefferson tried it, he was forced "to take a Mississippi River steamboat to Wheeling . . . then bump for twenty-four hours in a chilly stagecoach over rutted roads to Cumberland, Maryland, stopping every few hours for a meal while the horses were changed; then proceed by primitive train." When Harriet Beecher Stowe traveled from Cincinnati in 1834 to attend her brother Henry's graduation from Amherst College, the trip required ten days—and massive amounts of patience. She went "from Cincinnati to Toledo by stagecoach, to Buffalo by steamboat, to Albany by canalboat, arriving by stage in Amherst." Safe, reliable, and convenient it wasn't. People often stuck close to home simply because travel was expensive and complicated. Only the rich owned their own horses. To get from one place to another, most people walked. To get almost anywhere, you had to dearly want to go.

Just how bad were the roads? Stowe gives a vivid picture in *Uncle Tom's Cabin:* "In benighted regions of the west, where the mud is of unfathomable and sublime depth," she wrote of the regions through which Gatling would have traveled, "roads are made of round rough logs, arranged transversely side by side, and coated over in their pristine freshness with earth, turf, and whatsoever may come to hand, and then the rejoicing native calleth it a road, and straightaway essayeth to ride thereupon. In process of time, the rains wash off all the turf and grass aforesaid, move the logs hither and thither, in picturesque positions, up, down and crosswise, with diverse chasms and ruts of black mud intervening." And as more and more peopled headed west, bad roads became bad and crowded roads, increasing the danger as well as the inconvenience.

Improved roads would constitute "one of the magic keys to modernity," writes historian Paul Johnson. Roads. Byways. Trails. Turnpikes. Conduits. Without good roads, nothing could happen; with them, anything might.

When it came to bad roads, the United States and Great Britain faced the same bumpy, rutted challenge. And most everyone recognized the crucial need to change things. "Of all inventions," wrote British historian Thomas Babington Macaulay, "the alphabet and the printing press alone excepted, those inventions which abridge distance have done the most for the civilization of our species. Every improvement of the means of locomotion benefits mankind morally and intellectually as well materially."

John Loudon McAdam was the man who cracked the code. The wealthy Englishman was an amateur road expert, having peered intently at mile after mile of the atrocious British highway system—he traversed more than thirty thousand miles on his mission—to figure out precisely where and how often the wheels met the road surface, with what specific pressure and at exactly how many degrees of angle. His eureka moment came when he realized that small stones, not the large ones then in use, would constitute the best roads. "Every piece of stone put into a road which exceeds an inch in any of its dimensions," McAdam explained, "is mischievous. . . . A road made with small, broken stone, without mixture of earth, will be smooth, hard and durable." McAdam's thoroughfares were expensive to build, painstaking to maintain, but in the long run, they made a great difference for travelers and their worn-out backsides.

Until McAdam's roads took hold, though, people did the best they could. They rode horses or they balanced themselves on flatboats or they walked, or they spliced together journeys consisting of combinations of all three, and they exhibited tolerance, because they had no choice.

What with spills and snows and downpours and delays, it's a wonder that anyone went anywhere, but they did, on their way to seek new lives. Personal reinvention was to become a common theme in twentieth- and twenty-first-century America—behold the vast racks of paperback self-help books, the bountiful CD series advertised on late-night television for step-by-step transformation of body and soul and bank account—but in the middle of the nineteenth century, it still was a heady, fetching new idea. Herman Melville caught the mood along with the sea spray in his 1849 novel *White-Jacket:* "But our hearts are our best prayer-rooms, and the chaplains who can most help us are ourselves," his narrator declares, adding, "The Past is Dead, and has no resurrection, but the Future is endowed with such a life, that it lives to us even in anticipation. The Past is, in many things, the foe of mankind; the Future is, in all things, our friend. In the Past is no hope; the Future is both hope and fruition. The Past is the text-book of tyrants; the Future is the bible of the Free." Such rhapsodic sentiments created a powerful momentum in the new nation, in the energetic middle decades of the nineteenth century. "In the Past is no hope": This would be the land where the past didn't matter. The land of the future. The land of the second chance.

Eighteen forty-four, the year Richard Gatling left his family farm in North Carolina, was a momentous one. It was the year that the political crisis over the broad area known as Texas—and points west—was coming to its rough climax. The dour, taciturn, but determined Tennessean James K. Polk was elected president in 1844, pledging to annex Texas and acquire the Oregon territory to boot. The contest, in fact, was regarded by some as a de facto referendum on Western expansion. Lines were sharply drawn around competing camps. Polk, the Democrat, wanted to take Texas; his Whig rival, Henry Clay, did not. Clay and others, including John Quincy Adams, were afraid that the acquisition of Texas would not only rile up Mexico, just as the Oregon question was annoying the British, who claimed it as theirs, but also prove to be the tipping point in the perilously delicate balance between free and slave states. The election was a squeaker, but Polk prevailed by some thirty-eight thousand votes, and the West was now in play.

Newspaper editorial writers tried to outdo each other in spirited bombast and jingoistic invective. Led by John L. O'Sullivan, the essayist, explorer, originator of "manifest destiny," and later Confederate sympathizer, they took turns ratcheting up the rhetorical pressure for America's leaders to head west, to sweep in and take what God surely had ordained the nation to possess. In an 1845 editorial in the *New York Morning News,* a publication he had cofounded with Samuel J. Tilden, O'Sullivan declared: "Yes, more, more, more! . . . till our national destiny is fulfilled and . . . the whole boundless continent is ours."

O'Sullivan was matched, adjective for adjective, by an editorial that same year in the *Albany Argus*, urging Western expansion: "To live in such a splendid country . . . expands a man's views of everything in this world," because "their mighty rivers, their vast sea-like lakes, their noble and boundless prairies, and their magnificent forests afford objects which fill the mind to its utmost capacity and dilate the heart with greatness." In an editorial published September 25, 1845, James Gordon Bennett thundered in his *New York Herald:* "The patriotic impulses of the United States have been awakened to a fresh and greatly augmented vigor and enthusiasm of action. . . . The minds of men have been awakened to a clear conviction of the destiny of this great nation of freemen. . . . The pioneers of Anglo-Saxon civilization and Anglo-Saxon free institutions, now seek distant territories, stretching even to the shores of the Pacific; and the arms of the republic . . . must soon embrace the whole hemisphere, from the icy wilderness of the North to the most prolific regions of the smiling and prolific South."

He wasn't kidding about the "whole hemisphere" part. Some Americans had never taken their eyes off the far horizon. During the Revolutionary War, the Mississippi River had seemed like an acceptable western limit. By 1803, Thomas Jefferson's vision of the Rocky Mountains as the outermost western boundary of the United States struck many as reasonable. In the 1840s, nothing less than the Pacific Ocean would do.

An Indiana congressman, caught up in the giddy expansionist spirit, dubbed the process "the American multiplication table" during a House debate on the Oregon question: "Go to the West and see a young man with his mate of eighteen; and [after] a lapse of thirty years, visit him again, and instead of two, you will find twenty-two. That is what I call the American multiplication table. We are now twenty million strong; and how long, under this process of multiplication, will it take to cover the continent with our posterity, from the Isthmus of Darien to Behring's straits?"

Beneath all of these triumphalist, high-flying sentiments, though, were two somber truths on the ground: First, there was the intractable matter of the indigenous people who long had occupied these regions that Americans were so eager to settle. Second, many observers suspected that the push toward Western expansion was little more than a ploy to increase the number of slave states. Thus what sounded like innocent patriotic fervor actually was a canny gesture toward keeping slavery alive and well—and growing—in the United States.

No area, no matter how remote, could escape the slavery issue. The earliest settlers in St. Louis had been slave owners. Rapid economic growth meant that even greater numbers of slaves were needed, and needed fast. A traveler in 1818 recalled seeing flatboats, filled with slaves, riding the river from Louisville to St. Louis. Those slaves would suffer a fate even bleaker than the usual hell, because the slave owners in St. Louis were a notoriously mean lot. They believed they had to be. Illinois, which prohibited slavery, was tantalizingly close; slave owners in St. Louis were nervous because their valuable human property was close enough to freedom to smell it, to taste it. Hence those owners passed severe laws to keep slaves in line. Curfews for slaves were strictly enforced. William Clark, of the famous explorer tandem, dealt frequently in slaves from his St. Louis home.

So the West was a place of reinvention, of fresh vistas, but still it could not escape the familiar taint of slavery. It could not sidestep the central question of the time. No place could do so anymore. The nation was becoming too intertwined for that now.

Did most Americans believe in manifest destiny? Did they support the idea of expanding the nation just as far as it could go, marching westward right to the edge of the next ocean? Many very well may have, but perhaps not all of them supported it for selfish, acquisitive reasons. The earnestness that motivated so many other activities in nineteenth-century America, such as public lectures and the lyceum movement, also may have influenced attitudes toward Western expansion. Maybe it wasn't a land grab. Maybe it was a sense of mission. Maybe, as historian Frederick Merk has argued, manifest destiny did reflect the American spirit—but not in a pushy, headstrong way. Americans truly wanted to spread enlightenment, not just pile up land and get rich. Such a spirit was "idealistic, self-denying, hopeful of divine favor for national aspirations, though not sure of it," Merk wrote. It was "dedication to the enduring values of American civilization." Imperialism and expansionism were "never true reflections of the national spirit." That sense of mission—so redolent of the optimism and idealism of nineteenth-century values, so sincerely felt—would echo through the ensuing decades of American history, Merk wrote. "Mission appeared in the twentieth century as a national sense of responsibility for saving democracy in Europe. It was an important force . . . in inducing Congress in 1917 to vote entrance into the First World War. It did for the world, then, what it had not done for the newborn European republics in 1848." Thus a nation keen on giving second chances to individuals would grant the same golden favor to itself.

Could Americans truly have been so naïve? Could they really have believed that spreading their values through new territories in the West was an act of benevolence and generosity, not rapaciousness? They could. They could believe it in just the same way that the maker of the world's first working machine gun could believe, as he would always insist, that he created his gun to save lives, not end them. Yet that same Gatling gun would, shortly after its invention, become one of the weapons by which the West was swept nearly clean of its native population to make way for the surge of new settlers.

But what of that 1844 election, the outcome of which seemed, at least ostensibly, to indicate that most citizens favored Polk's nakedly expansionist plans? As Merk pointed out, most people actually opposed Polk. Most Americans didn't like the idea of a bullying land grab. This majority, though, split its vote between the Whig and the Liberty party candidates, both of whom were against Western expansion. So Polk and his ideas were backed by a minority of the voters. He slipped into office through a side door.

For Richard Gatling and an increasing number of his countrymen, heading west was a matter of financial prospects, not politics. The West meant new markets. Gatling, like so many others, had heard enthusiastic stories about possibilities in the West, about the huge rolling oceans of land in Indiana and Illinois and Missouri. There were reports such as this one, from a Scottish farmer who visited the United States in 1835 and made his way from Chicago to St. Louis—he alternated walking with riding in an oxcart, there being no other way to get there—and was astonished and moved: "I became fully sensible of the beauty and sublimity of the prairies. They embrace every texture of soil and outline of the surface, tall grass interspersed with flowering plants of every line. . . . Sometimes I found myself amidst of the area without trees or object of any kind within the range of vision . . . the surface clothed with interesting vegetation around me, appearing like a sea, suggesting ideas which I had not then the means of recording and which cannot be recalled." The Scotsman concluded: "The wide expanse appeared the gift of God to man for the exercise of his industry."

Land meant crops; crops meant farmers; farmers needed tools. And agriculture was what Gatling knew. Agriculture was what he'd been raised with. And agriculture was being swept up, as was everything else, in the nineteenth-century enthrallment with machinery. "The earth is a machine which yields almost gratuitous service to every application of intellect," wrote Ralph Waldo Emerson, himself the proud if mechanically inept owner of a ten-acre farm in

Concord, Massachusetts, in 1858. "Every plant is a manufacturer of soil. . . . The plant is all suction-pipe—imbibing from the ground by its root, from the air by its leaves, with all its might." Even plants could be viewed through the metaphor of industrialization.

Gatling had lots of ideas for farm implements. He envisioned steam-driven tractors and plows. The West was the ideal place to try out his notions. The West was the place everybody talked about for its size, its bracing newness, its rampant possibilities. It was changing quickly, the West was, it was becoming tamer and more sedate by the hour, it seemed, but still there was a wildness there, still an unknown element lurking amid the familiar. Railroads and steamboats were systematically linking cities, but still there was a raw, exotic feel to certain aspects of St. Louis, the river town to which Richard Gatling headed in 1844. St. Louis was becoming domesticated, steady, but there was still a roguish glint in its eye.

St. Louis was growing, it was assuming a place of real importance in the national economy, its businesses prospered—yet the city's edges still were ragged. The periphery glimmered mysteriously in the twilight. Just beyond the streets and fences, the land still was luscious and wild. A year after Richard Gatling arrived, an ad appeared in a St. Louis newspaper. It had been placed there by John Charles Frémont. The famous explorer was looking for men to join him on yet another expedition further west, the West of Colorado and California. By this time, St. Louis seemed like such a civilized place that Frémont wondered if he'd be able to rustle up enough adventurous souls to accompany him. There would be hardships aplenty. Was anybody still intrigued by the notion of unknown places, enough to give up a calm city life?

In the ad, Frémont had instructed interested candidates—whatever paltry few there might turn out to be—to meet at the Planters' Warehouse. When he arrived, he was stunned to be met by a large, noisy crowd, a crowd so enthused and unruly that Frémont had to move the group to a more spacious location. St. Louis had become a genuine city, all right, complete with "good schools & churches & some fine hotels," as a visitor at the time noted, but it retained a frontier flavor, too. As Frémont discovered, there were still a few Americans left who wanted to see the West, the real West, the vanishing West, at least once before they died.

Mark Twain, some two decades younger than Gatling, was growing up near St. Louis at just about the time when Gatling arrived. Twain watched the city change, watched its rapid absorption by civilization. All of his life, Twain would remember the beauty of this last, lost wilderness. He saw the raw and untamed part of St. Louis recede from view, disappearing in the dirty smoke of dozens of approaching steamboats, and while it was exhilarating, the change also induced an elegiac mood. "I can call back the solemn twilight and mystery of the deep woods," Twain wrote many years later, "the early smells, the faint odors of the wild flowers, the sheen of rain-washed foliage, the rattling clatter of when the wind shook the trees. . . . I can call back the prairie and its loneliness and peace, and a vast hawk hanging motionless in the sky. . . . I can see the woods in their autumn dress, the oaks purple, the hickories washed with gold." But it was dying, that paradise, and even though Twain loved the river with all his heart, he also knew that the river, as much as anything, was causing the death of a certain way of life, because the river meant progress, and progress had no use for oaks and hickories, except as fuel.

"That's the sadness," muses a character in Larry McMurtry's *The Wandering Hill*, a novel about travelers along the Missouri River in the 1830s. "There's not much time between first man and last man, between wild and settled. . . . We saw beaver who had never had to fear the trap, and buffalo that had never heard the sound of a gun. That was scarcely twenty years ago, and yet the beaver are almost gone and the buffalo will go next." As McMurtry's explorers also note, however, the West was still worth it. Its charms included the possibility of perpetual renewal. "In the freshness of the West," the narrator notes, "old ways could be peeled off."

When Lewis and Clark got their first glimpse of St. Louis in 1803, it was a small French Creole settlement ruled by Spain. It had been established four decades before by the French half-brothers Auguste and Pierre Chouteau, who had traveled up the Mississippi from New Orleans until they found a spot that appealed to them. Here, the Missouri River split off from the Mississippi. They marked it with a few knife gouges in a tree. Trees, being plentiful, were natural boundary stakes. St. Louis itself, you could say, was carved out of a tree, established by pushing away a portion of the hickory, pecan, oak, black walnut, ash, and cottonwood trees that grew in beautiful profusion, interrupted only by hemp. Hemp, too, was prevalent. Yet by the early 1800s, there were already unmistakable signs of civilization's heavy step: Buffalo herds were alarmingly thinned. In just a few short years more, the onrush of steamboats from places such as New Orleans and Louisville would cover the sun with the black smoke from their hungry, wood-burning engines. A newspaper editorial in 1823 complained that the grimy atmosphere in St. Louis was "so dense as to render it necessary to use candles at midday. Rain in passing through it has been so discolored as to stain everything with which it came into contact, like ink." Elk and antelope still roved nearby, but their numbers dropped with each passing year.

By the time Gatling arrived in 1844, St. Louis had become a full-fledged city. It had also become, literally and symbolically, the threshold to the rest of the West. It was the jumping-off point. After St. Louis, forests gradually subsided into prairies; prairies segued into plains. The waters of the Missouri River—increasingly, that was the way new residents arrived in the city, first on flatboats and later on steamboats—could be tricky and rough and had even given Lewis and Clark some anxious moments, but river transport was unstoppable, dangerous or no.

St. Louis would go from fewer than seventy-five thousand people when Gatling arrived to almost half a million less than a decade later. It was the dividing line between the city and the wilderness, between the refined and the chaotic—or, as Frederick Jackson Turner would put it in his seminal 1893 essay, "The Significance of the Frontier in American History," it was "the meeting point between savagery and civilization." St. Louis featured elements of both. It was genteel and unpredictable. It was safe and perilous. It was calm and frenzied. Its first bank opened in 1816—but promptly went bust. Not until 1829 would the citizens have a bank upon which they could depend. A wide-open, anything-can-happen perkiness still could be teased out of the settled-seeming atmosphere. It wouldn't last. On July 27, 1817, Zebulon Pike brought the first steamboat to St. Louis. Two years later, the riverfront was thronged with steamboats. The city never looked back.

It wasn't just the West, of course. The West only dramatized, in its rugged and vivid and volatile way, what was true of the nation as a whole: America was the land of the second chance, where success depended on a surfeit of effort and pluck, not a fancy coat of arms. Success came to those who worked and dreamed. Henry Adams called the urban versions of these upstarts "new men": They were immigrants, many of them, and they had dodged difficult circumstances and painful blunders. They had shed old entanglements. Now they were remaking themselves, in a nation that allowed—even encouraged and celebrated—such second acts. The businessmen who helped create the financial colossus of nineteenth-century America were men such as John Jacob Astor and John D. Rockefeller; one a threadbare immigrant, the other the desperately poor son of a con man. Both rose to immense financial heights. They were men such as Alexander Stewart, another immigrant, who created, from a pocketful of Irish lace, a mammoth store in nineteenth-century New York. The era "demanded a new type of man—a man with ten times the endurance, energy, will and mind of the old type," Adams wrote.

Visiting the United States in 1845, Prussian historian Frederick von Raumer marveled over the nation's "elastic vigor of youth," further noting that "the poetry of the Americans lies not in the past but in the future. We Europeans go back in sentiment through the twilight of ages, that lose themselves in night; the Americans go forward through the morning dawn to day!" Turner would echo that in his manifesto, describing it as nothing less than the very definition of the national identity: "This perennial rebirth, this fluidity of American life . . . furnish the forces dominating the American character." America, the second-chance land, had the space to accommodate any number of fresh starts in new places. And like so many others, twenty-six-year-old Richard Gatling didn't want his father's life.

The prospects for a life such as his father's did not, in any case, seem especially bright just then. In 1844, the year Richard Gatling left North Carolina, cotton prices dropped to the lowest point they'd ever been in the country's history. Admittedly, the Midwest was faring little better than the South, as prices for corn and hogs also had been decimated by a long depression that began in the early 1840s and hung on, hung on, ruining many farmers. But if things are bad everywhere, you go anywhere: Why not? There seemed little point in sticking around. So people headed west, cobbling together new lives. "New communities are springing up in the West every month, on whom the past has but little hold," wrote Irish journalist E. L. Godkin, after touring mid-nineteenth-century America. "They have no history, and no traditions. . . . The West, in short, has inherited nothing, and so far from regretting this, it glories in it. One of the most marked results of that great sense of power by which it is pervaded, is its strong tendency to live in the future, to neglect the past." There were legions of people such as Mark Twain's father, Virginia-born, West-obsessed, always ready to pick up, move on, try again. And again. "Like thousands of other ambitious young men of his time and place," wrote a Twain biographer, "he [Twain's father] moved westward with his growing family again and again—five times in a dozen years—betting that first one frontier settlement and then another would flourish and he with it."

America was the place to do it. And those who went west "held under their hammers a thousand miles of mineral country with all its riddles to solve, and its stores of possible wealth to mark," Adams stated. "They felt the future in their hands."

John Sutter surely did. The German merchant had left Europe in 1834 and come to America, where second chances seemed to lie thick on the ground. He needed one. He'd made a mess of things back in Switzerland, where he'd tried his hand at business, and he left behind towering debts and sputtering creditors. In the United States, Sutter created a highly successful trading post in California. It was on his land, in the magical year of 1848, that gold was discovered. In America, anything could happen, it seemed. Why, you could pluck gold out of the very streams.

In 1842, Samuel Colt was a bankrupt gunmaker who had made all the wrong decisions when it came to running a business. He had lost thousands of dollars that family and friends had staked him. He had angered and disappointed just about everybody who had ever believed in him, ever taken a chance on him. Everything he touched had gone rotten. Yet a mere four years later, Colt had reestablished himself, and his revolvers were tumbling off the assembly line in a shiny new factory in Hartford, Connecticut, with impressive and lucrative regularity.

There were few practical impediments to the individual renaissance. There were not many ways for your past to dog your steps, because the record—what record there might be—was difficult to check. Others had to take what you told them on faith. People could move beyond the disgraceful things they'd done. They could outrun a shabby past. They could reassemble themselves, bit by bit, and emerge as entirely new individuals, shining and hopeful. In the middle of the nineteenth century, the United States was still small enough to enable such transformations. It was still a handcrafted world. There was an artisan quality to the population as well as the economic base: People, like things, were unique and distinctive, created one at a time, knowable as discrete entities. That would change, of course. As the population soared throughout the nineteenth century, and as products increasingly were made on assembly lines in factories, rather than by hand in barns and backyards, people and things became ubiquitous. They could be dealt with only in the aggregate. The land of the second chance would give way to a country of interchangeable parts, interchangeable people.

For all of the times that this ability to slip the shackles of the past ended up letting a con man off the hook or enabling a criminal to hoodwink a gullible acquaintance, those episodes were balanced by a life such as Ulysses S. Grant's. He was the consummate second-chance man. The eventual hero of the Civil War was, only a few years before the conflict began, a washed-up failure who moved about in a cloud of misery and despair. Had America been another kind of country at the time, had it not been a place in which lives could have second acts, Grant would have had no opportunity to lead the Union troops.

In the twenty-first century, if a president were to appoint a man with Grant's spotty background to such an important post, the outcry would be deafening. Most likely, he or she never would be allowed to serve. But in a second-chance land such as nineteenth-century America, Grant got his shot.

Just how devastatingly unimpressive was his early biography? In the late 1850s, Grant appeared to have no prospects. After distinguished service in the Mexican War, the West Point graduate had seemed to fall away from his future, like bark sloughing off a tree. Drink surely contributed. Living on a squalid sixty-acre farm outside St. Louis with his family, Grant applied for job after job, to no avail. He wrote to his father back in Ohio, asking for a loan to tide him over. No reply.

As a last resort, the ragged-looking, unkempt Grant peddled firewood on the streets of St. Louis. Sometimes a soldier he'd known would come along, and do a quick double take: Could this be Grant, the valiant officer? No. It couldn't be. Could it?

"Great God, Grant, what are you doing?" asked a startled former comrade from the Mexican conflict, one sad afternoon.

Answered the somber Grant: "I am solving the problem of poverty."

In time, he would redeem himself. He would recreate himself. Just as his country would be forced to go through the hell of a civil war to remake itself, to decide what kind of country it was going to be—progressive or archaic, free or slave—so Grant, too, would go through heat and flame. He would be transfigured. He would emerge purified, in a sense, but certainly different. Later, after a scandal-flecked presidency and the diagnosis of terminal cancer, Grant would have to do it all yet again, the reinvention, the new start, and because of where he lived, he was able to accomplish that, for whatever time might be left to him. "Humiliated, bankrupt and voiceless, on the very threshold of death, sleeping at night sitting up in a chair as if he were still in the field and could not risk losing touch with developments, he relived his old campaigns," wrote Edmund Wilson, describing the ordeal Grant suffered as the general wrote his memoirs. To stay where one was, mired in dismay and disappointment, was optional. Fatalism had no place in this world. Nothing was ordained. Man and country could recapitulate themselves, after desperate trial and long effort. Everything depended on the next gesture, the second wind. The second chance.

In 1844, Richard Gatling settled down to work as a clerk in a dry goods store owned by a man named William Adriance. That same year, the telegraph was patented. Amazing changes were in the air. Yet down on the ground, where people lived and worked each day, it was still a rough and bedraggled time. A filthy one, too. Major cities had no public sewer systems. For the distasteful but necessary duty of disposing of human wastes, people employed chamber pots or, if they had the yard space, outhouses. Away from home, they simply utilized a nearby ditch or culvert. The idea of a daily shower or bath would have struck them as laughable, not to mention impossible. America in the 1840s, one historian noted, was "a dirty place inhabited by dirty people." Streets were blurred with mud. Horse manure was a hazard affecting most every step. People wore hats, boots, and long cloaks not for style, but generally to protect themselves from dust and weather. Women, if they could afford to, fought personal foulness with excessive amounts of perfume, which, instead of canceling the stench, only added a peculiar new layer of unpleasantness to it. Fingernails were framed in grime. The backs of most people's necks were gray with sweat and filth. Public buildings were awash in spittoons, and they were not for show. Sanitation and personal hygiene were exotic concepts. Many wells were routinely contaminated. Not surprisingly, life expectancy for the average person was just under forty.

These were primitive days for the young country. But the future was there, too; the future was already writing its name in the sky, a sky turned black by the smoke from wood-chewing steamboat engines. The future was carried in the piercing yell of the steamboat whistle, the kind that made hearts jump and dogs howl. In private homes, objects such as sofas and carpets—unheard of just a decade before, except among the very rich—were showing up. Candles had given way to oil lamps. Chairs, no longer strictly handmade, became affordable for average families, because those chairs were turned out on assembly lines, produced in large quantities. The idea of buying a chair, instead of making one, must have been astonishing at first. Ready-made clothes, too, could be purchased now at a reasonable price by ordinary people.

By 1850, the railroad seemed to be everywhere. Iron rails bit into the countryside. On steamboats, propellers replaced paddle wheels. On ocean vessels, iron hulls forced wooden ones into retirement. A nation of isolated settlements, tossed as if at random across a sprawling landscape of forest and field, flung like dice on green velvet, was becoming a place of integrated markets. Of spiraling interdependencies. Instead of producing things to sell to your neighbors, or to people in the nearest town, or to people in the next town over, you produced things to sell to the rest of the country, and to the world. And the way those things were produced also was changing. Industrialization was spreading. Rural self-sufficiency was retreating.

In the time he had off from working at the dry goods store, Richard Gatling supervised the making of several of his patented planters, and then he sold those planters to farmers in the area. They were a big hit. Gatling's invention was so successful that after a year, he was able to quit his day job and embark on a full-time career peddling the agricultural devices he had dreamed up.

Gatling's first job away from his father's farm, when he set up shop in a country store in rural North Carolina, had been at a crossroads. Now, he found himself at another one. His country was in the same position. The 1840s were the crucial decade during which the new intersected with the old. The machine began to supplant human muscle. It was not an absolutely clean break, of course; the essence of a crossroads is that elements are still present from both directions. Covered wagons and stagecoaches still rocked and swayed along rutted roads, even as the occasional locomotive staggered past on twin ribbons of rail. Rafts and flatboats and keelboats still shared the river with newfangled steamboats. And St. Louis itself—vibrant and brash, easily the most important port anywhere north of New Orleans—was gradually but steadily losing ground to Chicago, its "glorious rival," as one newspaper put it in the early 1850s.

The reason was simple: Rail transport was leaping ahead of the river routes. In 1840, the nation had 2,700 miles of rails; in 1850, it had 8,683 miles; in 1860, the astonishing total was 30,283 miles. More trains were going to Chicago than to St. Louis. Chicago had allied itself with the North, with the cities of the East Coast and their constant commerce by rail; St. Louis was still identified with New Orleans and points south, with river traffic. "The new gateway sat not at the confluence of rivers on the route to New Orleans," writes historian William J. Cronon, "but at the Lake Michigan railhead on the route to New York." St. Louis had bet on rivers in more ways than one; by 1840, steamboat construction had itself become an important part of the city's economy. There was plenty of timber and, even more significantly, plenty of foundries and ironworks and machine shops by then to produce engines and boilers for steamboats.

The path to the future was heading in another direction now, away from St. Louis. River transport was a long way from being obsolete—in the 1850s and 1860s, steamboats still left their scrambled wake up and down the Mississippi and the Missouri and the Ohio, night and day, and would for decades hence—but change was coming. The country's population was shifting and recalibrating itself as new spaces opened up, as new outposts took hold, and a young man named Richard Gatling, toting his seed planter, took to the rivers and the roads to show it off and sell it at state fairs and agricultural demonstrations, to talk to anybody who'd listen.

• • •

"TO FARMERS AND SPECULATORS!" the poster begins, in thick, serif-rich capital letters. "Offered for sale, THE PATENT RIGHT to counties & states of the greatest Invention of modern times, namely, A SEED SOWING HARROW, which sows and covers the seed at the same time, thereby saving much labor and time to the Agriculturalist."

Then the poster, which Gatling distributed in St. Louis and used on sales trips up and down the river, settles itself after the typographical hysteria of tall capital letters and continues: "The above machine sows the wheat in narrow streaks or drills, which is far preferable to any other way. Experience has proven, beyond all doubt, that wheat or any other kind of small grain, thus planted, will produce a great deal more than when planted the ordinary way. In England, where the object of the Agriculturalist is to raise all he can to the acre, wheat is invariably sown in drills, and rice planters, in the south, all drill the rice, which adds at least one fourth to the crop." It concludes: "A model of this cheap and valuable invention can be seen at any time. Persons are requested to call and examine the machine for themselves. Terms—A credit of Six Months, with improved endorsed paper."

So before machine guns it was seed planters, harrows, farm implements. Practical devices. Gatling was still almost a decade and a half away from inventing the world's first efficient multiple-firing weapon. Once the gun made him famous, many people were surprised to find out about his early patents for agricultural devices. Given the simple, logical configuration and operational ease of the Gatling gun, they naturally assumed its inventor had long been a gunmaker. They thought he had surely spent many years perfecting his armaments designs, as had Samuel Colt. But no. The seed planters and hemp brakes, the drills and the steam-driven tractors, actually were at the heart of Gatling's creative interests. The gun was an anomaly. The gun was the big surprise.

CHAPTER FOUR

"Drunkards, Dandies & Loafers"

The river persuades you into its way of thinking. Motion is a state of optimism and everything about the river lives in the present.

—Russell Celyn Jones

In 1843, a rather priggish and chronically observant twenty-one-year-old man from Adams, New York, decided to step out and see the world. He couldn't step very far before he had to board a steamboat, because America's rivers were the first national highways, spreading out across the land in a crooked network like the spidery cracks in a dropped mirror.

Henry Benjamin Whipple went around the American horn—or what horn there was in a new nation in the 1840s. First he went south, then northwest, and finally northeast again. Part of his journey was by stagecoach, part by train, but mostly it was by steamboat.

Years later, Whipple would serve as a powerful and distinguished Episcopal bishop in Minnesota, but this was 1843, and he was simply a traveler with a good eye and a busy pen and just enough prickly disdain for his countrymen to keep his opinions interesting. Time and distance have made the world of Western steamboats seem somewhat quaint, picturesque, but Whipple was an irked and snippy eyewitness, elbowing his way through the crowds that gathered at dockside, coughing at the smoke from the bossy steam engines and dodging the piled-up carts and the skipping children.

Leaving St. Louis for Cincinnati by steamboat was, he recounted, a cacophonous carnival: "You are besieged until the last moment by newspaper sellers, by book pedlars, candy women, fruiters & others who are as importunate as possible to buy! Buy! . . . [T]he heavy steamboat tolls for you all an adieu. Whiz, puff, puff, and back we go into the stream, a revolution or two of the wheel and at the signal the cannon is fired and away our noble boat rushes on her course like a steed under whip and spur."

Once on board, Whipple noted, the wildness never abated: "Instead of a quiet and a reasonable number of passengers, we had over 300 and such a motley crew of mortals I will wager never before was seen in one boat. . . . Men smoked in the cabin, spit showers of tobacco juice on the carpet, talked loud, laughed heartily." The steamboat, Whipple wrote, was jammed with "tall Kentuckians and short pigmies [sic], drunkards, dandies & loafers. Oh! such a crew!"

This was the world Richard Gatling entered, too, when he left St. Louis in 1845. He spent the next year or so traveling on steamboats from river town to river town, showing off the agricultural implements he had invented. Gatling went from St. Louis to Cincinnati to Pittsburgh and points in between. His life deftly intersected with the era of steamboats in the West, a relatively brief time—railroads would come along shortly and steal the steamboats' thunder—that was nonetheless startling and remarkable for passengers and beholders alike.

Steamboats first made the West what it was, linking towns along the river, and then linking that river to another river, and connecting its towns, too. The commerce that steamboats enabled was important, yes, but there was something else. There was also just the fantastic, unbelievable *fact* of steamboats, the raucous reality of the big white smoking behemoths, two or sometimes three decks high, that were "rushing down the Mississippi, as on the wings of the wind," as one observer commented, "bearing speculators, merchants, dandies, fine ladies, every thing real and every thing affected, in the form of humanity, with pianos, and stocks of novels, and cards, and dice, and flirting, and love-making, and drinking, and champagne, and on the deck, perhaps, three hundred fellows, who have seen alligators, and neither fear whiskey, nor gunpowder."

Steamboats were a crucial economic force in the early West and an important means of transportation. They were able to shave days and weeks off journeys that formerly could only have been accomplished by stagecoach or on horseback. But steamboats also were extraordinarily dangerous. They were slapped together quickly and cheaply; in the 1840s, more than 150 new steamboats were built each year, and the average lifespan for a Western steamboat was less than four years. Goaded by pilots eager to stay on schedule, their rickety boilers often were pushed past a safe operating capacity. Explosions were frequent and appallingly injurious to hapless passengers. Another hazard came from submerged rocks which, when hit, could quickly sink the boat. The rivers were winding and tricky, with channels that frequently shifted. The currents could be treacherously swift.

The deadliest aspect of steamboats, however, had nothing to do with steam or currents. It had nothing to do with explosions or collisions. The worst peril associated with steamboats was that they inadvertently gave a helping hand to the deadliest disease in human history: smallpox.

As long as people lived far apart from each other, as they did in North America in the early 1800s—indeed, as had been the case across most of the planet until then—the smallpox virus could not gain a foothold. But the steamboat brought together many people who would otherwise have remained separated, dwelling in areas that were distant and isolated. The virus requires a human host. It can only move from person to person. It can only multiply inside the cells of the afflicted. It cannot survive outside an infected person.

In the mid-1840s, steamboats began to link populations as never before. And along with the pianos and novels and cards and dice and champagne, the steamboats carried something infinitely less welcome: the smallpox virus. Among the victims of the smallpox epidemic that ravaged the Ohio River Valley in the winter of 1845–46—one of several outbreaks that would occur throughout the 1840s—was a handsome, ambitious young inventor named Richard Gatling.

It was a close call. Smallpox did not kill him, but it changed him forever, inside and out. The virus carved scars on Gatling's face, the telltale flecks and chinks that people in the nineteenth century came to know all too well.

A young William James survived a bout with smallpox. Cornelia Seward, infant daughter of William Seward, secretary of state in the Lincoln administration, died of it in 1837. When it came to smallpox, some people lived, and some did not, according to the inscrutable whim of fate. By the middle of the century outbreaks had become increasingly common, as transportation networks spread and people rubbed elbows as never before. Yellow fever and cholera and other diseases also were carried aboard steamboats, but the chief worry, the worst scourge, was smallpox. It killed faster and more horrifically than any other. Although the virus itself is infinitesimally tiny—a thousand smallpox viruses could straddle a single human hair—its death toll has easily surpassed that of any other infectious pathogen in history. A wary Henry Whipple, watching from the deck of the steamboat as Wheeling, West Virginia—at that point, still Virginia—came into view during his 1844 journey, decided to stay put: "I did not go much about town as the small pox is here & prevailing to a very considerable extent." Cities such as Cincinnati and Richmond began to post yellow flags along the riverbank. It was a way of warning steamboat passengers that a smallpox epidemic was rampant. Ironically, of course, there was every chance that an earlier steamboat had actually brought the virus ashore in the first place.

The smallpox virus requires a population of about two hundred thousand people who live within a radius of a two-week journey of one another. Otherwise, without new human hosts to keep its life cycle going, the virus dies when its victim dies. As long as the United States was a place of widely scattered settlements and small isolated farms, as long as it remained primarily a rural agrarian land, the virus was not a substantial threat. But then came the steamboats. Then came the exciting and lucrative river traffic. Then came all of those majestic and colorful vessels hustling up and down the Western rivers, creating a vast, interlocking market—and a hospitable home for the smallpox virus. Among the unsuspecting Native American population, the disease "swept up that river like a blaze," a character observes in a Larry McMurtry novel about the 1840s West. "People couldn't die much quicker if you shot them. . . . Most of the Missouri Valley is just a land of ghosts now." Even if they didn't know what to call this malady, they knew what it did to them, and they saw how it followed the waterways, twisting and winding as the riverbank did, like a hunter keenly focused on the track of a scent. Wherever the river went, the awful thing that brought death went there, too. It was, McMurtry wrote, "the great sickness that seemed to hang over the river." Progress always had a price.

The smallpox scars on his face, Gatling would later say, were what first persuaded him to grow a beard. He kept the beard for the rest of his life. But the internal changes were what mattered most. The things people couldn't see.

He had been on his way from Cincinnati to Pittsburgh in the winter of 1845–46. The steamboat was locked in the ice along the

Ohio River between the two cities, forced to remain idle for almost two weeks. Inside his cabin, Gatling sweated and fretted, his scalp, face and body covered with small, painful pustules. Noninfected passengers were allowed to disembark at Wheeling, but Gatling and the other sufferers had to stay on board. Even afflicted with the less virulent and usually nonlethal form of the disease, known as varioloid rash, Gatling would have been in severe discomfort. Pustules scabbed and broke open, releasing the pervasive and repellent smell of pus. Scabs dropped from a victim's skin in a constant sprinkle, leaving behind the ugly, disfiguring scars. Yet the scabs themselves were the true ugliness; they contained the virus and would lie in wait, wherever they fell, until the next human host came along. Those scabs accounted for the swift spread of smallpox among unwitting crowds on the decks of steamboats, or in any area where people were pressed in tight for prolonged periods of time.

Had Gatling made it to shore in Pittsburgh, he still would have found only minimal relief. The first hospital in the city—indeed, in all of western Pennsylvania—did not open until 1847. Some cities operated what were called "pest houses" to provide shelter for smallpox victims; these were little more than shacks, thrown together to get the unsightly ill off the main thoroughfares. They were filthy and crowded. Often they were little more than resting places for the grim interval between desperate illness and death.

Shortly after his recovery, Gatling learned that his seventeen-year-old sister, Martha Sarah Gatling, had died back in North Carolina. Two other sisters, Mary Anne and Caroline, also had died young, at twenty-five years and twenty-nine days old. Life was a fragile and uncertain thing in those days. For a driven young man such as Gatling, the idea of making the most of one's time, whatever its duration, now was paramount. His own close call, and the deaths of his sisters, apparently persuaded him to make an abrupt change in his plans. In 1847, he enrolled in the Indiana Medical College in LaPorte, Indiana.

Compliments of R. J. Gatling
Hartford, May 1st 1893.

Richard Jordan Gatling with the bulldog version of the Gatling gun, May 1, 1893. *(Courtesy of E. Frank Stephenson, Jr.)*

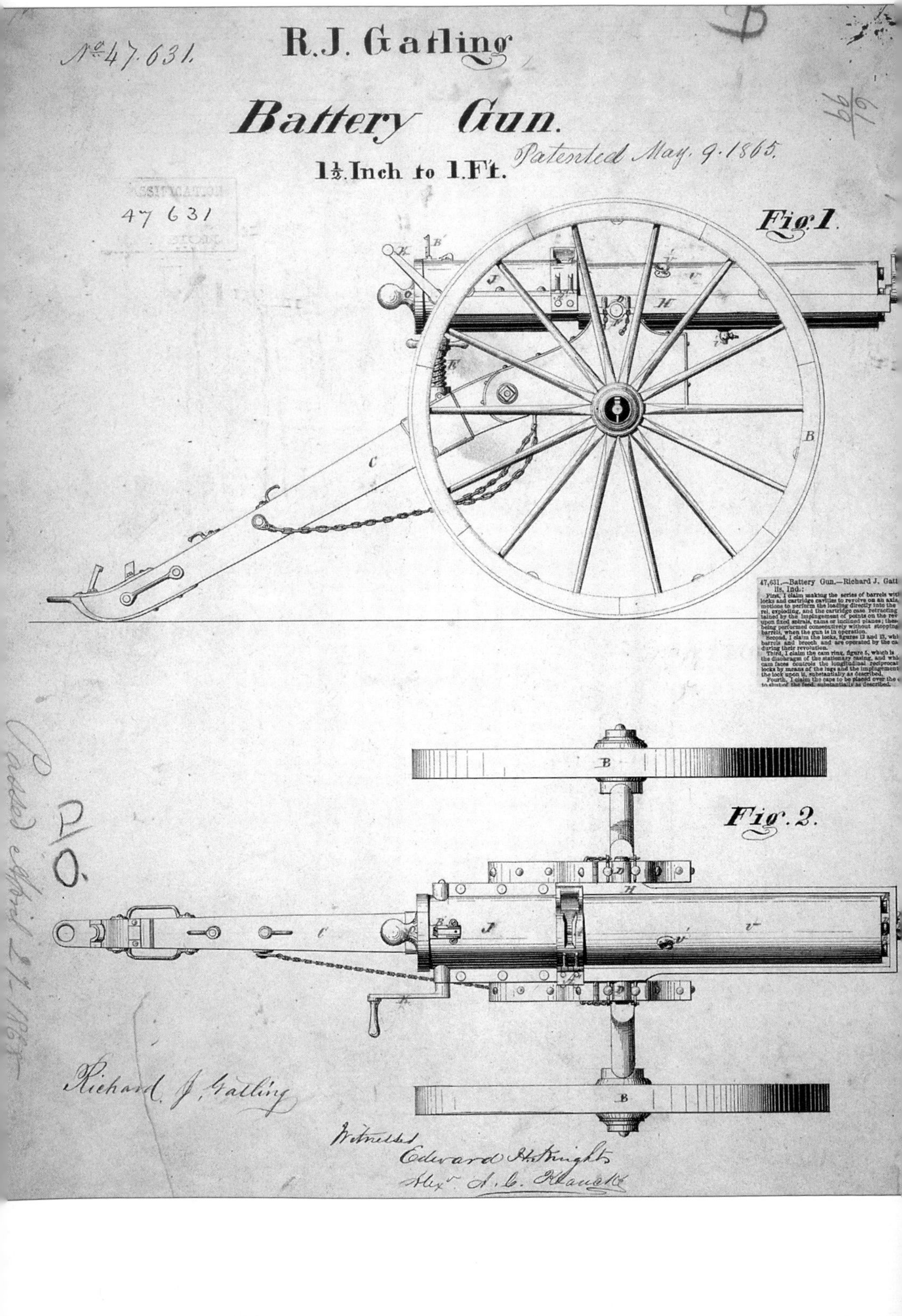

No. 47.631.

R. J. Gatling

Battery Gun.

Patented May. 9. 1865.

1½ Inch to 1 Ft.

Fig. 1.

47,631.—Battery Gun.—Richard J. Gatt
lls, Ind.:

First, I claim making the series of barrels wit
locks and cartridge cavities to revolve on an axis,
motions to perform the loading directly into the
rel, exploding, and the cartridge case retracting
tained by the impingement of points on the rev
upon fixed spirals, cams or inclined planes; thes
being performed consecutively without stopping
barrels, when the gun is in operation.

Second, I claim the locks, figures 12 and 13, whi
barrels and breech and are operated by the ca
during their revolution.

Third, I claim the cam ring, figure 5, which is
the diaphragm of the stationary casing, and whi
cam faces controls the longitudinal reciprocat
locks by means of the lugs and the impingement
the lock upon it, substantially as described.

Fourth, I claim the caps to be placed over the
to shut of the feed, substantially as described.

Fig. 2.

Richard J. Gatling

Witnesses
Edward H. Knight
Alex. A. C. Klauck

OPPOSITE: Patent for 1865 version of the Gatling gun, May 9, 1865.
(Courtesy of the National Archives; Patent No. 47.631)

Richard Gatling in 1854 in Indianapolis.
(Courtesy of E. Frank Stephenson, Jr.)

Jemima Sanders in 1854, on the eve of her marriage to Gatling.
(Courtesy of E. Frank Stephenson, Jr.)

The transport steamboat *Chattanooga.* Steamboats speeded up travel, but were prone to explosions and helped spread smallpox. Gatling traveled on such steamboats in the late 1840s.
(Courtesy of the National Archives; Photo No. B-672)

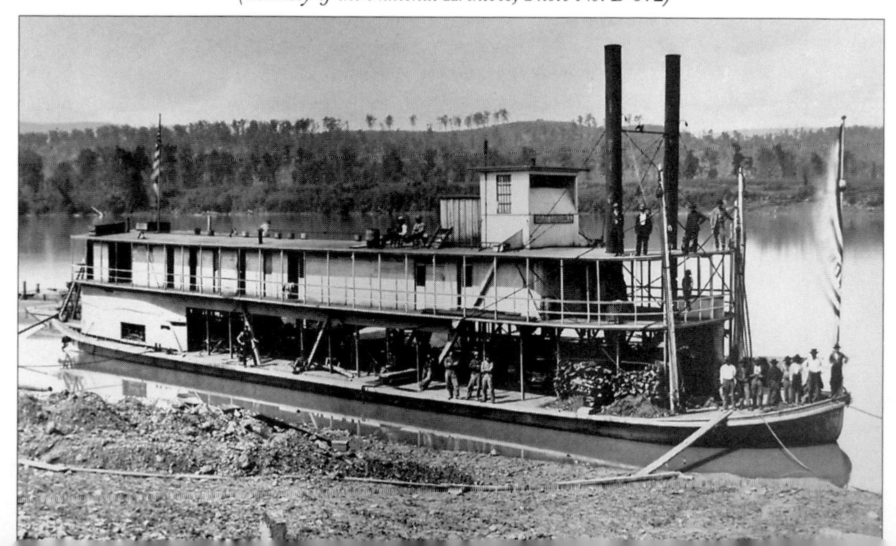

Governor O. P. Morton of Indiana, who pushed hard for President Lincoln to buy Gatling guns. *(Courtesy of the National Archives; Photo No. BA-1040)*

A Gatling gun at the Washington Arsenal, in the late 1860s. The Potomac River is in the background. *(Courtesy of the National Archives; Photo No. 156-AA-69)*

Trials of Gatling guns at the Washington Arsenal, May 8, 1866. *(Courtesy of the National Archives; Photo No. 156-AA-68)*

General James Wolfe Ripley, chief of ordnance during the Civil War. He hated the idea of machine guns and kept the Union Army from adopting Gatling's weapon. *(Courtesy of the National Archives; Photo No. BA-203)*

Lincoln kept a close eye on troops and weapons. Here in Antietam, in 1862, he discusscs strategy. *(Courtesy of the National Archives; Photo No. B-2933)*

General Ben Butler. An early champion of Gatling guns, he bought and fired them during the siege of Petersburg. *(Courtesy of the National Archives; Photo No. B-3751)*

OPPOSITE: A war president: Abraham Lincoln in the field with (*right*) General George McClellan and (*left*) Allen Pinkerton. *(Courtesy of the National Archives; Photo No. 111-B-634A)*

A steel engraving of the world-famous inventor Richard J. Gatling, in the late ninteenth century.
(Courtesy of E. Frank Stephenson, Jr.)

A gleaming row of Gatlings, hot off the factory floor at the Colt Patent Firearms Co. in Hartford, Connecticut.
(Courtesy of E. Frank Stephenson, Jr.)

Gatling's famous neighbor in Hartford, Mark Twain. Of the city, Twain wrote, "You do not know what beauty is if you have not been here."
(Courtesy of the National Archives; Photo No. BA-1085))

Buffalo Bill Cody, pictured here with Sitting Bull, stoked the exploited imagery of the Old West: Indians and Gatling guns.

(Courtesy of the Library of Congress, ID cph 3a22279)

The bulldog Gatling—nicknamed the police Gatling—in action in the late 1890s. Gatling guns were used by law enforcement officers against striking workers and unruly mobs. *(Courtesy of E. Frank Stephenson, Jr.)*

Promoting the Gatling gun abroad in the late nineteenth century, this publicity shot referred to the smaller weapon as a "camel gun." *(Courtesy of E. Frank Stephenson, Jr.)*

Hiram Maxim, whose 1881 machine gun edged out the Gatling, also invented more weapons; here, he shows off his ill-fated flying machine. *(Courtesy of the Library of Congress, call number LC-B2- 4064-2)*

The family man: Richard Gatling with his son, Richard Henry Gatling, and grandson, Addison Barnes Gatling, in New York in 1901.

(Courtesy of E. Frank Stephenson, Jr.)

"It's the Gatlings, men, our Gatlings!": Theodore Roosevelt, who uttered the famous cry as he stormed San Juan Hill during the Spanish-American War, is pictured here in 1899, spoiling for a fight.
(Courtesy of the Library of Congress, ID cph 3a40275)

A Gatling gun crew in the Philippines in 1899.
(Courtesy of the National Archives; Photo No. 111-SC-98352, Box 751)

Special units, such as this one, were trained to handle Gatling guns in the Spanish-American War. *(Courtesy of E. Frank Stephenson, Jr.)*

By 1900, when this photo was taken in New York, Richard Gatling had suffered numerous financial setbacks, but still pressed on with new inventions. *(Courtesy of E. Frank Stephenson, Jr.)*

The twenty-nine-year-old Gatling had already been a farmer, a teacher, a merchant, an inventor, and a salesman. Now he was determined to become a doctor.

Everywhere one looked, it seemed as if another new technology was bringing people and places and ideas together, canceling the isolation that formerly had been the rule in a sprawling nation of sparse population. Distance was erased. Time was compressed. In 1844, there came the stringing of the first telegraph line; in 1866, the draping of the first successful transatlantic cable; in 1869, the driving of the spike at Promontory Point, Utah, that completed the Pacific Railroad and created the first transcontinental railway. A truly mass culture was dawning. When the hammer smacked the spike on May 10, 1869, "the entire nation was now connected in one great electrical circuit," writes historian David Howard Bain. And like steamboats and their complicated cargoes of both shiny new pianos and deadly smallpox viruses, these new technologies brought good things and bad things, improvements and setbacks, treasures and peril. Before railroads, steamboats had demonstrated that new technologies always come at a price—yet it was a price that Americans were eager to pay. "Steamboats took fire, bumped into one another, hit rocks and sank," a river watcher wrote, "hit sand-bars and stuck for hours or maybe days, were liable to be tied up by low water or carried into the woods by high, ran into logs, wrecks, and keel-boats, and withal prospered, multiplied, transported the moving millions, and performed an indispensable function to society and civilization."

Among the most horrific consequences of steamboat travel were the explosions. Most of these occurred on the Western rivers, where the steamboats were flimsier, the captains less cautious, the speeds higher, and the confluence of all of these elements sometimes catastrophic. Fires, snags, collisions, and damage from storms were one thing, but an explosion from a bursting boiler on a steamboat was quite another. When a steamboat christened the *Clipper* exploded on the Mississippi in 1843, an account in the *Louisiana Chronicle* did not hold back on the gory details: "Machinery, vast fragments of the boilers, huge beams of timber, furniture and human beings in every degree of mutilation, were alike shot up perpendicularly many hundred fathoms in the air. . . . The hapless victims were scalded, crushed, torn, mangled and scattered in every possible direction. . . . The 2nd engineer [was] thrown 150 or 200 yards through the roof and gable end of a house, into the back yard against the fence—one arm torn off and fragments of his carcass scattered over the trees."

In February of 1858, some twelve miles below Baton Rouge, the calm of a Sunday morning was riven by an explosion aboard the *Princess.* A half century later, a man who had chanced to come upon the aftermath was able to recall excruciating details that had clearly haunted his sleep for decades: "I can never forget the agonizing sight that met my gaze at the little cottage filled with the suffering and dying officers and passengers," wrote John E. Rowland. "We ministered to their needs in every possible way, with sweet oil, mattresses and medicines." To little avail. Until the federal government began to require steamboat inspections in 1852, reliable numbers for accidents were not kept; based on newspaper accounts, however, between 1846 and 1848 there were at least twenty-eight steamboat explosions on Western rivers, killing 259 people.

A decade later, the threat remained. Eighteen fifty-nine was the year that Henry Clemens, nineteen-year-old brother of Mark Twain, died along with two hundred others when the boilers on the steamboat *Pennsylvania* blew up on the way from New Orleans to St. Louis. The injured were taken to Memphis, where a devastated Twain watched his beloved sibling expire. "For forty-eight hours I labored at the bedside of my poor burned and bruised but uncomplaining brother," Twain wrote to his sister, "and then the star of my hope went out and left me in the gloom of despair."

Lives were not the only cost. Until the 1850s and the switch to coal, the engines burned wood; each day, steamboats used up the equivalent of seventy square miles of the nation's forests. Ohio had sacrificed half its woods to peckish steamboat engines by the middle of the nineteenth century.

Still, there was no stopping the steamboats. The rivers were irresistible. The Mississippi, some twenty-three hundred miles long, was four miles across at its widest point and at its skinniest, a mere thirty feet. Its depths ranged from two hundred feet to three feet. It held more water than all the rivers in Europe combined. By 1859, the thousand-plus steamboats that sputtered up and down the Mississippi and its tributaries were hauling more goods at any given time than all of America's ships on all of the world's oceans. The Ohio River, on which Richard Gatling spent most of his time, was no slouch, either: "No man will ever forget his first view of the Ohio," wrote an observer named Charles Fenno Hoffman, as he stood on its banks near Wheeling. "The clear majestic tide, the fertile islands on the bosom, the bold and towering heights opposite . . . and the forest-crowned headlands, above and below, round which the river sweeps away . . ." The steamboats that frequented these waters created a motley, hurly-burly hodgepodge, as Whipple noted, especially at mealtimes: "What a rush, one grand race, and woe to the luckless wight who should stop in his course, he might well expect to be crushed to death—and then such a clatter of knives and forks and table ware, such screaming for waiters . . . such an exhibition of muscle & nerve as men entered with all their powers into the fame of knife and fork. It was worse than a second Babel . . . food was *bolted* as I have never seen before. . . . At night the cabin looked like one great hospital, so close were they stowed away. . . . As for the quality of our food, Oh! Never mention it."

Western steamboats, upon which Whipple was traveling, valued cargo above people; that is why the accommodations were primitive, the decks crowded, the crew preoccupied. In the East, where passenger travel was the focus, the steamboats were usually swanky and luxurious. The very best of these "floating palaces" were apt to have mahogany staircases, crystal chandeliers, silk curtains, and mirrors and carpets in staterooms. Steamboats everywhere, though, drew a crowd. "Orange-women and news-boys assail you at every step," wrote a traveler in 1866, confirming that things were little changed from Whipple's day two decades earlier, "whilst the hoarse voices of escaping steam-waste, and the discordant tintinnabulation of a score of bells, hurry on the laggards by warning of the near approach of the hour of departure. . . . Several bells suddenly cease, when from different slips, steamboats covered with passengers will shoot out like racehorses from their grooms . . . and begin the voyage with wonderful speed."

Twain had seen the worst that steamboats and rivers could do to human beings, and he still couldn't stay away. He earned his riverboat pilot's certificate the year after his brother's death. The river "had a new story to tell every day," Twain believed. For the author, one of his biographers says, the Mississippi was "a wondrous interruption in that landlocked prairie, a fabulous anomaly, a constant *event*, motion in the midst of stasis." There was shared, among the men who followed the river trade, an obsession for the liquid immensity that surrounded them, an emotion that was part love, part fear, part awe. Such a feeling was inspired by few other professions. "It was a great, absorbing, dominating, vital, and to the outsiders, unexplainable passion for the river," claimed one observer. "To say that it was fondness for the river or a liking for the river—that would mean nothing. What they felt was more like a lover rejoicing in his first love, only this was immune from satiety and did not change. . . . It was a romance that never grew tiresome, that never lost its savor. . . . For him no life had taste or tang but the life of the river. . . . Let any man with any touch of the poetic or emotional in his being make so much as three trips on the river and he was done for. From that time the spell had him fast." Like so much else in this pivotal age, steamboats brought equal parts danger and opportunity. Railroads were shortly to become the transportation icons of the nineteenth century, the symbols of swift change and inventive fire; but steamboats were first. They were the first to link remote outposts with established cities, the first to begin binding a new nation into an economically complex whole. They were the means by which men such as Richard Gatling got where they needed to go. They were beautiful, and exciting, and sometimes deadly, and the peril surely added to the excitement.

"You have undertaken no ordinary task. You may feel discouraged at its magnitude, and disposed to seek some simpler and humbler pursuit; but do not falter too soon. To a chill succeeds a fever; on a momentary revulsion of feeling, from the sudden appearance of danger, there follows a reaction, which nerves the arm and achieves a victory." These were among the ringing and inspirational words that Dr. Daniel Drake delivered on November 5, 1849, to the students at the opening term of the Ohio Medical College in Cincinnati. After beginning his studies at the Indiana Medical College in LaPorte, Gatling had transferred to the Ohio school.

"The morning sun may not be able to send his beams through the mists, and fogs, and clouds, which hang over us in autumn," Dr. Drake continued, "but he rises higher, and shines with fierce and warmer rays, until they gradually melt away."

By the end of 1849, Gatling completed his coursework, or at least enough of it to call himself "Doctor" with a relatively clear conscience. Apparently he did not take the trouble to actually graduate, but in medicine's early days, when degrees were negligible, that didn't really matter. The surest route to becoming a doctor in the 1840s? Start calling yourself "Doctor." Indeed, one of the most famous "doctors" of the nineteenth century, the man who made a fortune selling patent medicine, was Lucius S. Comstock of New York. His credentials consisted entirely of having pasted "M.D." to the end of his name. Later, for good measure, Comstock threw in a bogus law degree, too.

Gatling may have been in medical school, but he wasn't ready to renounce his earlier passion. He continued to conceive of new mechanical ideas, sketching, tinkering, daydreaming, even as he attended classes. And there was, on the board of trustees for the Ohio Medical College, a man who was definitely simpatico with Gatling, a prominent Cincinnati businessman named Miles Greenwood. Greenwood, too, believed in technology; he ran a successful iron foundry in this increasingly important Ohio River town. Gatling and Greenwood may very well have first met each other at this time. They would reunite in late 1862, when Gatling needed a factory to make his new weapon and Greenwood had just the place.

During the Civil War, intense suspicion would come to settle upon Gatling for picking Cincinnati as the place to make his gun. Gatling was Southern-born, and surely, some thought, he must have harbored secret affection for the old homeplace and its rebellious ways. Perhaps he was a Copperhead—the derisive name given to Northerners who wished for a Confederate victory. If that weren't so, if he were a true-blue Union man, his detractors would mutter, why had he selected Cincinnati, of all cities, as the birthplace for his new weapon? Cincinnati, which seethed with secret Southern sympathizers? Cincinnati, which was directly across the Ohio River from the roiling, contested state of Kentucky? Throughout the early part of the war, Kentucky teetered on the edge of secession, so strong were its slave-holding roots.

There were other foundries in other cities. Why Cincinnati—unless he was, as the rumors claimed, a Copperhead who hoped that

Southern guerrillas might swoop in and swipe the odds-changing new weapon, without Gatling's having to bear the responsibility or the consequences of having given it to them? But this theory missed the key fact that Gatling's association with Cincinnati and Miles Greenwood went back many years, back to the late 1840s, when a bright young man sat at a desk in a lecture hall and scribbled copious notes on subjects such as chemistry, anatomy, and surgery.

He used a small butterscotch-colored notebook, leather, rubbed-looking, soon frayed at the edges. It fit in his palm. He packed the lined pages with a hasty, spiky-looking script. On the inside back cover, Richard Gatling practiced his signature, trying it one way, then crossing that out and trying it another way. And then he added the pale penciled proclamation: 1849.

"Gun shot wound," was the underlined topic of one page. "The ball in entering the flesh frequently changes direction. If larger joints & bones are much injured, amputation is often necessary—give stimulants, narcotics." The lecturer must have quickly moved on: "Tetinus [sic] is lockjaw," Gatling dutifully wrote. "1st symptom is stiffness of muscles about the neck—opium and alcohol in large doses is best remedy." In another section of notes, Gatling wrote "Cholera," underlined it, and added, "If brought on by the state of the atmosphere—it is not contagious." The remedy? "Opium & stimulants," he wrote. There were long lists ("Magnesia," "Lime," "Sulphates") and simple declarative statements ("Iodine—it is good for scroffula [sic]"). Obstetrics clearly was part of the curriculum: "Child is viable at 7 months," Gatling noted, adding "At full term weighs generally 7 or 8 [lbs], is about 18 inches long."

The profession Gatling was seeking to enter in 1849 was still in thrall to ancient theories, still finding its way within the unsettling exhilaration of scientific breakthroughs. The American Medical

Association had been formed in 1847, but medical schools were not professionalized until the 1870s. The first full-time professor of medicine in an American institution of higher learning, Robley Dunglison at the University of Virginia, had only been appointed in 1825. Medicine was still very much an apprentice system, with aspiring doctors hooking up with physicians who had existing—and, ideally, lucrative—practices. Patients didn't scan the office walls for diplomas. The venerable English system of "humours" in the body, liquids that encouraged diseases when they tilted out of balance, so influential in Elizabethan days, still held sway in early-nineteenth-century America.

Mostly, medicine remained a matter of folk remedies and hopeful guesses, not laboratory work, not professional practice. A lot of the people who called themselves "physicians" had no training in the sciences whatsoever. Medical schools had few academic standards; witness the fact that Richard Gatling, with very little formal schooling, was admitted without a fuss. In the main, doctors were "self-instructed and self-certified."

In 1849, the Ohio Medical College was less than three decades old. There was ample reason for Harvard University President Charles Eliot to complain at the time: "An American physician may be, and often is, a coarse and uncultivated person, devoid of intellectual interests outside of his calling, and quite unable to either speak or write his mother tongue with accuracy." If a medical student showed up for a few classes and passed a nominal exam, he could obtain a state license and call himself "Doctor." Only after the Civil War did subjects such as pathology and bacteriology become requirements in medical schools. In 1893, professors at Johns Hopkins University took the then-outlandish step of stressing laboratory work.

A medical "profession" was slow in taking hold. This was the age, after all, of patent medicine, when many people relied on pills sold in bottles that sported enticingly colorful labels. In the late

1830s, the most popular remedy in the country was Dr. Morse's Indian Root Pills, sent from New York City to points west, wherever people had aches and pains and complaints of a vague but persistent sort. The first patent medicine in the United States, Dr. Samuel Lee's Bilious Pills, was a tremendous success almost from the moment it was introduced in the 1790s; before his death in 1805, Lee could revel in the fact that his pills were found in homes far and wide. This was a hazard of the era of the amateur: In an important field such as medicine, the lack of uniform professional standards and a system of credentialing meant that anybody could sell anything to everybody, with virtually no oversight. Yet the same freewheeling, unregulated world that enabled Richard Gatling to call himself "Doctor" after a few medical school courses was the same one that would, a few years hence, enable a self-taught industrial engineer named Richard Gatling to invent a new type of gun and gain the financial backing to have it manufactured—all without a formal degree in the field or verifiable apprenticeship in the profession.

Medicine at this time was as much a matter of salesmanship as of science. In the 1840s and 1850s, traveling salesmen supplied most of the patent medicine, stopping in at drugstores and general merchandise stores to restock their shelves. In the early 1850s, patent medicine companies began to give away almanacs with their products, the better to cozy up to customers. The Comstock Company, the nation's best-known purveyor of patent medicine, boasted in advertising material that its sales empire stretched "from the Maritime provinces to the Mississippi Valley, and from Ontario—or Canada West—to the Gulf." Patent medicines were, in effect, harbingers of the mass-marketing concept. Department stores did not appear in the United States until the 1880s; national chains were unknown before 1900. Yet the extravagant promises that accompanied the all-purpose cures were part of the same pitch one could hear in Maine or Missouri. People seemed to lap up those pitches, especially when they came with fantastical back stories. Dr. Morse was a fictional creation, but his made-up biography was infinitely consoling to those who relied upon his pills to alleviate their pains. There was an entertaining, adventure-filled yarn about how he had come across his spectacular product, and the good doctor often showed up in advertising illustrations with his "healthy blooming family." Dr. Morse, customers were reliably informed, "was the first man to establish that all diseases arise from the impurity of the blood." And the claim that he'd discovered his health-restoring medicines among the Native American population was a satisfying notion. Indians—who symbolized earthy, "natural" living, eschewing fancier, more sophisticated medical knowledge—popped up in a great many patent medicine brochures.

"Dr. Larzetti's Juno Cordial or Procreative Elixir," read the enticing label on a product sold in 1844, is "a certain remedy in all cases of Impotency, Barrenness, Fluor Albus, Difficult or painful Menstruation, Incontinence of Urine, and all diseases arising from debilitation of the system where an impulse or restorative is required." A decade later, the Comstock Company was still going strong. Its sales brochure listed such miracles as Kline's Tooth Ache Drops, Judson's Worm Tea, and Dr. Chilton's Fever and Ache Pills. Following in subsequent years were Kingsland's Chlorinated Tablets for All Throat Diseases, Dr. Howard's Seven Spices for all Digestive Disorders, Dr. Howard's Blood Builder for Brain and Body, and Dr. Chapman Hall's Canker and Dyspepsia Cure.

The Blood Builder pills were valuable for especially delicate personal issues. According to an advertising circular: "They have an especial action (through the blood) upon the SEXUAL ORGANS of both Men and Women. It is a well recognized fact that upon the healthy activity of the sexual apparatus depend the mental and physical well-being of every person come to adult years. It is that which gives the rosy blush to the cheek, and the soft light to the eye of the maiden. The elastic step, the ringing laugh, and the strong right arm of the youth, own the same mainspring." After only a few doses of Dr. Howard's pills, "the impoverished Blood is enriched. The shattered nervous forces are restored. Vigor returns. Youth is recalled. Decay routed. The bloom of health again mantles the faded cheek." Some entrepreneurs, no matter what their core business was, could not resist the lure of selling unregulated medicines as well. During his first transactions as a fur trader in New York in 1800, an enterprising German immigrant named John Jacob Astor also peddled an early version of Viagra: an herb guaranteed to enhance male virility.

As late as the 1880s and 1890s, hundreds of patent medicines still were being made and distributed throughout the country. A directory for druggists, published in 1895, listed some fifteen hundred such products. The claims for these so-called remedies were cheerfully fraudulent, concocted by con men with a flair for dramatic ad campaigns. The public, then as now, believed what it wanted to believe.

Yet the changes that were occurring everywhere else in the country, the improvements in transportation and communication, the upgrades in education, in hygiene, gradually would come to the medical profession as well. "Experience, down to the present hour, has shown," Dr. Drake added that fall day in 1849, in his address to Gatling and his classmates, "that every medical school in the West has suffered more or less of a revolution in its forming stage; the inevitable effect, perhaps, of a new and unsettled state of society." American medical schools, like the nation itself, were on the move. More knowledge among the general public and consumer protection laws finally began to pick away at the patent medicine business.

Nonetheless, it was a slow process. Even late into the nineteenth century, medical schools weren't fully professionalized. Standards and curricula varied wildly from institution to institution. Scientific breakthroughs were routinely upending knowledge of the body and its functioning, requiring a head-to-toe rethinking of previously inviolate certainties. The word "scientist" was not even coined until the 1830s. And speaking of words, just what was the term "doctor" supposed to signify, anyway? Nobody was quite sure. Naturally, quacks purveying patent medicines were more than happy to rush in and fill that murky void with magic elixirs and fabulous potions. Dr. Drake's institution suffered through a scandal in 1878, involving the medical-school necessity that most distressed the general public: the procurement of cadavers. It seems that Senator John Scott Harrison, son of President William Henry Harrison and father of President Benjamin Harrison, passed away and was buried. His son, in the meantime, received a report that the corpse of a recently deceased family friend had been filched from its grave and smuggled to the Ohio Medical College, where students were preparing to cut into it for their anatomy studies. Arriving in Cincinnati, Harrison's son made a startling discovery: It was the body of John Scott Harrison, son and father of presidents, not the family friend, that lay on the dissecting table.

As medical schools struggled to guide the discipline toward science and away from folklore, the grip of the patent medicine business on the public imagination proved exceedingly hard to pry loose. The Federal Food and Drug Act passed in 1906, but in 1918, Dr. Morse's pills still did a brisk business as sure-fire cures for "biliousness; dyspepsia; constipation; sick headache; scrofula; kidney disease; liver complaint; jaundice; piles; dysentery; colds; boils; malarial fever; flatulency; foul breath; eczema; gravel; worms; female complaints; rheumatism; neuraligia; La Grippe; palpitation; nervousness." Habits died hard. You could also, if you were so afflicted and so inclined, still pick up a bottle or two of Dr. McNair's Acoustic Oil, Dr. Sphon's Head Ache Remedy, or Dr. Connol's Gonorrhea Mixture.

• • •

For "Dr." Gatling—and he would never be shy about whipping out the quasilegitimate honorific if he thought it might help sales of his inventions—the world was both large and small. He had lived in two major cities, St. Louis and Cincinnati, but he was also part of the constantly shifting Venn diagram of which nineteenth-century history seems peculiarly constituted, the accidental overlaps, those pivotal if unwilled interactions that just seemed to happen in this era, over and over again. Along with Miles Greenwood, who would figure in Gatling's life as an inventor, Gatling may have crossed paths with someone else during his medical school days: Reuben Samuel, later to become the stepfather of Jesse James when he married James' widowed mother, Zerelda, in 1855. Gatling and Samuel could very well have known each other, because Samuel graduated from the Ohio Medical College in 1850, which means he was there in 1849, Gatling's last year at the college.

No records confirm an acquaintance, but the two men were at the same school in the same city at the same time, and surely passed each other on the street, in the lecture hall, in a tavern, or in a theater. The world was compact enough then to accommodate such random brushes, such casual overlaps. It was still an intimate place. The population in the United States was still small enough to make such happenstance possible. Thus the future inventor of the world's first working machine gun and the future stepfather of the world's best-known outlaw could very well have exchanged pleasantries in a cloakroom one morning or nodded briefly at each other across a crowded restaurant. Between the Revolutionary War and the Civil War, the country was still young and petite, and the great crisscrossings—the times when two figures, destined to make their marks in different ways, passed one another in their early days, unaware, of course, of the contingencies that later would bind them in the history books—could repeatedly occur. Abraham Lincoln and Jefferson Davis were born in shambling cabins in western Kentucky within a hundred miles of each other, only a year apart. "Two Kentuckians: one went this way, the other went that way, and of the men who followed them 600,000 lost their lives," noted Bruce Catton.

As a boy, the father of Ulysses S. Grant once boarded with a clan named Brown, Grant recorded in his memoirs. Orphaned early, the senior Grant became a tanner and "lived in the family of Mr. [Owen] Brown, the father of John Brown. . . . I have often heard my father speak of John Brown, particularly since the events of Harpers Ferry. Brown was a boy when they lived in the same house, but he knew him afterwards, and regarded him as a man of great purity of character, of high moral and physical courage, but a fanatic and extremist in whatever he advocated." Thus the father of the Union general who was to finally win the war lived, for a time, alongside the man whose violent actions were to help precipitate that war.

Lines, links, tangles: The stepmother of Lew Wallace, Civil War general and author of *Ben-Hur* (1880), was the younger sister of Jemima Sanders, wife of Richard Gatling. Wallace later served on the special jury that decided the fate of the conspirators in the assassination of President Lincoln, which was death by hanging.

Circles, angles, connections: Such is the essence of history in this period, the chance encounters and inadvertent echoes and incidental weavings and twinned shadings, the places where things scrape and merge and separate, unbeknownst. So much is unpredicted and unplanned. So much is mere occurrence—not script, not impulse, not desire. "It seems," a dying Ulysses S. Grant wrote to his doctor, "that man's destiny in this world is quite as much a mystery as it is likely to be in the next." Much dangles at the mercy of historical happenstance. Long lines of lives that look isolated and purposeful are revealed, when a little distance is obtained, to be snarled tangles, to be chronic meanderings toward a random rendezvous with other lives. By the early 1850s, Richard Gatling was a resident of Indianapolis, moving steadily toward the resounding events and great personalities of the age. Some of these intersections were willed, hoped for, and some of them were not.

Why did Gatling go to Indianapolis? It made sense in many ways. Indianapolis was a good place for an inventor-turned-physician who was shortly to turn back into an inventor again, and a businessman as well. The city embraced and enabled such turns and tumbles. The city itself had a kind of tentative, freewheeling feel. Indiana had become a state in 1816, and Indianapolis was laid out five years later, its stakes set up along the White River to take advantage of a steamboat trade that, alas, was never to materialize. The river was too shallow. Indianapolis had to retool itself for another kind of future. In 1825 it was named the state capital, owing mainly to its location in the center of the state. It had yet to shed a kind of backwoods, primitive aura. When Henry Ward Beecher arrived in 1839 to take over as pastor of the Second Presbyterian Church, he was unimpressed: "The whole city was given over to politics and money making," he grumbled. Those priorities might have disappointed a minister, but they suited an entrepreneur such as Gatling right down to the ground. River commerce hadn't exactly worked out, but by God, railroad building surely would.

Like so much of the West, Indianapolis's rise seemed rocketlike. In 1822, just a single two-story house interrupted its horizon. By 1823, it had almost one hundred families and its own newspaper, even though the streets were still muddy, and the place retained a rough and uncouth feel, an unpeeled frontier rawness. The population surged past a thousand in 1827. A year and a half later, a visitor wrote that Indianapolis "begins to look like a town . . . [with] ten stores, six taverns, a court-house . . . and many fine houses." By 1850, when Gatling arrived with intentions to settle down, the city had 8,091 residents. Only a decade before, the population had topped out at 2,692. Big things were ahead.

He seems never to have practiced medicine in Indianapolis, but he did most everything else. Gatling reinvested the substantial profits he made from manufacturers who had purchased the rights to make his seed planter. He began to deal in real estate, betting on the spread of the railroads. In 1851, his harrow was displayed at the Crystal Palace Exhibition in London, to much acclaim.

In a tintype profile made in Indianapolis in 1854, Gatling looks trim and gallant, with a neatly razored beard, eyes small enough to tempt the description "beady," and a slab of straight dark hair that drops beyond his elegantly upraised collar. There is a kind of twinkle in his eye, too, a sense that he's both inside and outside this picture; a man of his time but a man hovering just past the present moment, too, a man who knows he'll one day be looking back on all of this, having left his stamp, having made his mark. Portraits of other individuals from the same era sometimes have a solemn, aggrieved-looking ambivalence to the faces, an uncomfortable wariness, amid the extravagant whiskers and fancy high collars. Not a bit of that here. Gatling looks almost merry. His mouth is touched with a faint but findable smile.

He sat for this picture shortly before his marriage to a round-faced, big-eyed beauty named Jemima Sanders. Gatling's wife-to-be was the youngest daughter of Dr. John H. Sanders, a prominent Indianapolis physician. Jemima's sister Zerelda was married to David Wallace, the Indiana governor whose sons included William Wallace and Lew Wallace. Zerelda was also a well-known advocate of women's rights. If Gatling's intention was to rise even higher in Indiana business and political circles, he could not have selected a better entrée than marriage into the Sanders clan. They were a distinguished and accomplished family with an array of well-known friends. They were heavily involved in business, politics, the military, even the arts: After his Civil War service, Lew Wallace would go on to write *Ben-Hur,* one of the top-selling novels of the nineteenth century, achieving more fame as an author than he ever achieved as a Union general at Shiloh.

By entering this golden circle, Gatling made himself part of the Indianapolis elite. He moved easily, gracefully, in their ranks, even though he was an outsider. When it came time eight years later to organize the Gatling Gun Company, Gatling would call upon these acquaintances to stand with him, to invest in his company, to serve as corporate officers. Gatling sought out local stalwarts such as W. H. Talbott and John Love, the latter a general in the Civil War; these were important men who knew a thing or two about the business world, men who had taken the measure of this Richard Jordan Gatling and found him worthy. Gatling later would add Edgar T. Welles, son of Gideon Welles, secretary of the navy in the Lincoln administration, to the top ranks of the Gatling Gun Company. Gatling had no problem, that is, finding men of notable reputation who believed in him and his invention.

The thirty-six-year-old Richard Gatling and the seventeen-year-old Jemima Sanders were married on October 24, 1854, at the home of Jemima's other sister, Mrs. Mary Elizabeth Duncan. The bride's father had died seven months before the ceremony.

Gatling was a married man now, a business leader, a vital part of a thriving city. He and his wife settled into a roomy house on the southwest corner of Michigan and Delaware streets. As is surely the case with most families, their lives were a blend of joy and shadow, of blessing and grief. Two children, William and Mary, died in infancy; a third, five-year-old Mary, died in 1860. Their fourth child, Ida, turned two that same year, and her health—God willing—seemed sound. Later, after the family had moved to Hartford, Connecticut, two more children would come along: Richard Henry Gatling in 1870, and in 1872, Robert Barnes Gatling.

By 1860 business was going well for Gatling, now forty-two; his railroad investments in Indianapolis and elsewhere were holding their own. His financial dealings in the growing city were paying off. Gatling was a rich man. But the old fire still burned. The hunger. The familiar passion. Gatling found that he could not tear himself away from inventing things. In 1860 alone, a decade after he had moved to Indianapolis, he obtained five patents: a rotary plow; a cultivator for cotton plants; a lath-making machine; an improved hemp brake; and a gearing machine. Something else, however, something entirely different, lay just out of sight in his imagination, waiting. But not for long.

In the summer of 1861, just months after Fort Sumter and the nation's slide into all-out war, Richard Gatling had an idea. It wasn't entirely new; its vague contours had suggested themselves to him years ago, when he formulated the mechanics for his seed planter. Fed by a gravity-driven hopper, the seeds dropped, one by one, into the furrow. Gatling couldn't get that process out of his mind: its rotating simplicity, its smooth mechanical perfection.

On many days during that tense and uncertain first summer of the war, Gatling found himself down by the Indianapolis train depot. He saw the volunteer regiments board the trains and depart, in a gray haze of smoke and good wishes, and then he would see regiments return, their numbers drastically thinned. As stretcher after stretcher was pulled off the cars, Gatling watched the sick and wounded men on those stretchers as they twisted and writhed or, even more ominously, lay still. More than two-thirds of the casualties in this war would be from disease, not combat. It was worse than a simple human tragedy, he thought; it was a grotesque and shameful waste.

Gatling talked it over with an acquaintance named Benjamin Harrison, another rising Indianapolis businessman. Harrison recently had been put in charge of the Indiana Volunteers, and he shared Gatling's outrage and dismay. Harrison wanted to give his men their best chance at survival—and that chance just might rest, the future president speculated to his inventor friend, in superior weaponry. Weaponry that required fewer men to operate it. Weaponry so machinelike in its efficiency that it would discourage an opponent into surrender, thereby truncating war itself.

Gatling had attended a lecture earlier that year given by an arms maker from Ohio. The visitor had explained the benefits of a breech-loading cannon that could fire quickly and continuously. Now Gatling had everything he needed: the basic mechanical design, embodied in his seed planter; the moral imperative, supplied by the memory of the dead and ailing soldiers as they arrived at the Indianapolis train station; and the commercial impulse, which arose as a possibly awkward but completely predictable consequence of the realization that this might well be a drawn-out, expensive affair. It might last years, not months, despite all the breezy hypothesizing at the outset. And long wars meant large profits for gunmakers.

Simple designs were best, Gatling knew. Some inventors tried to complicate things, tried to trick out their devices with fancy-looking frills. But when he sat down on a succession of those anxious summer days and sketched out his idea for a machine gun—or battery gun, as it was then called—Gatling summoned what he knew about gravity, about force, about recoil, and the result was a design so exquisite, so fundamentally sound, that it is still used by gunmakers today. It was a superb feat of engineering, accomplished by a completely self-taught mechanical engineer. It was the work of an amateur, in an age when being an amateur lacked its later stigma: It meant you saw things with a fresh eye, untainted by rules and traditions. It meant you possessed an uncluttered imagination.

Six barrels—which were rifled, meaning that helical grooves were carved along the insides, enabling the bullet to grip the grooves as it exited, and consequently fly straighter and faster—were fixed around a central axis. Each barrel had, at the breech end, a breech block and a striker. Attached to the closed end of the barrel was a percussion cap—a type of primer that was filled with fulminate of mercury, first developed in 1830 to replace the clumsy, unreliable flintlock system. Ammunition for this first Gatling gun consisted of a paper cartridge loaded with powder and a .58 caliber bullet. The cartridge was placed in a steel tube; the tube was inserted into the feed hopper—the same kind of gravity-fed hopper that Gatling had used with his seed planter. This hopper, however, fed cartridges instead of seeds into each subsequent barrel.

A crank at the side rotated the barrels. The strikers turned with the barrels; as each striker moved, it hit the percussion cap, thereby igniting the powder and creating the charge that fired the bullet. The spent cartridges dropped out of the bottom of the gun and could be reused. With each revolution of the crank, six rounds were fired.

Only the very first Gatling gun employed paper cartridges; thereafter, the gun was redesigned to shoot metal .58 caliber rimfire bullets. Indeed, in the years to come, Gatling would tinker tirelessly to perfect his gun's design, adapting it for customers' particular needs, incorporating advances such as smokeless powder and, in 1890, an electric trigger. But the fact that this fledgling version worked as well as it did—firing some two hundred rounds a minute in initial trials—is absolutely remarkable, given Gatling's dependence on the primitive armaments technology of the day. Paper cartridges were handmade, lacking the consistency and uniformity of mass-produced metal bullets. The gun's efficiency at this time, many armaments historians agree, was testimony to the phenomenal mechanical genius of Richard Gatling.

No one is quite sure where the initial prototype of the Gatling gun was made. A factory in Indianapolis may have been the site, although there is also a persistent tale that a metalsmith in Freeport, a small community thirty-five miles southeast of the state capital, did the honors.

When Gatling sat down to invent his gun, it was as if all the rivers in his life suddenly were flowing in the same direction. Everything before that moment, all that he had done and dreamed of doing, all of his travels and his sorrows and his joys and his restless midnight musings, seemed to coalesce in six barrels and a central shaft and a hand crank.

In later years, a weary Gatling, a battered Gatling, a man who had been hollowed out by business failures and betrayals by colleagues, would say, *No, no,* the Gatling gun was not really his life's work, not at all. He would stop just short of wishing aloud, when he was old and frail and near death, that he had never invented the gun in the first place. He would insist it had cost him more money than he had ever made from it. If only, he would say, he had stuck with seed drills and steam-driven plows. His last nine patents had nothing to do with armaments. A century later, Mikhail Kalashnikov, inventor of the Avtomat Kalashnikova 1947, or AK-47, would say much the same thing: "I wish I had invented a lawnmower," Kalashnikov told an interviewer in 2002. The AK-47, like the Gatling gun, made mass killing easier and cheaper than ever before.

As Gatling's gun first took shape on the page, though, regrets and second thoughts were still a long way off. A kind of restless, golden energy took hold of him, pushing him forward. Less than a year and a half after the first shot at Fort Sumter, Richard Gatling had invented, designed, and tested a revolutionary new weapon and signed a contract for its manufacture. The year 1862 was a remarkable one in the history of human destructiveness; it was also the year Alfred Nobel, working in a borrowed laboratory in St. Petersburg, mixed glycerin alcohol with nitric and sulfuric acid, perfecting the formula for nitroglycerin. Death on a massive scale was now at the world's fingertips.

When O. P. Morton, the feisty Indiana governor, was told of the great new gun this Gatling fellow had created, he, too, got busy. He dashed off letters; he hectored bureaucrats and military men; he cajoled and he sweet-talked. There was no time to lose. A terrible war, stubborn and bloody and black, had the country in its grip, and the governor believed he might have found a way to curtail it. He had discovered a secret weapon—a weapon which, if the pushy and forthright Morton had his way, wouldn't stay secret much longer.

CHAPTER FIVE

The Spaces Between the Bullets

Confederates who beheld it [the Gatling Gun] said, "The Yankees have a gun you load on Monday and shoot all the rest of the week."

—Bruce Catton

Abraham Lincoln was hooked. He loved mechanical things, he was enthralled by technology, and the first time the sixteenth president saw a machine gun, he couldn't get enough of it. He was already fascinated by steamboats and railroads and telegraphs. And a new kind of weapon was really just an offshoot of those innovations. Another turn of the wheel. Lincoln would watch a machine operate and subsequently wonder how, and why, it did precisely what it did, and then he'd replay the process backward in his mind, reverse-engineering what he beheld. He had a nineteenth-century kind of mind, which is to say he observed things not with a back-away awe but with a get-up-close curiosity: What makes it *go*?

Once the Civil War began, Lincoln took an earnest and abiding interest in guns, especially a multiple-firing weapon nicknamed the coffee-mill gun—a clumsy precursor to the Gatling gun, an inadequate gun that eventually would prove to be a large thorn in the side of Richard Gatling. Lincoln, though, saw the coffee-mill gun before the Gatling gun was more than a gleam in Richard Gatling's eye, and the president liked what he saw. Lincoln was always intrigued by firearms. And when he peered long and hard at some of the funny-looking versions brought around for his inspection, his fascination was obvious. The physical signs of intellectual absorption—crimped brow, squinting eyes, pursed mouth—must have deepened the grooves in a face already scored and cross-hatched by perpetual fretfulness. These latest lines, though, came by way of fascination, not woe, and maybe didn't cut quite so deep. This president whose face was "so awful ugly it becomes beautiful," as an eyewitness named Walt Whitman saw it, was a man who appreciated tangible things as well as abstract concepts; beneath the dry exterior of a brilliant brooder, Whitman added, was "a fountain of first-class practical telling wisdom." Lincoln's father had been a talented carpenter, a man who built churches and wagons and corner cupboards with finesse; the future president is thought to have helped his father on such projects, which meant that he knew how to use his hands. He knew his way around hammers and saws and pegs. Lincoln was practical as well as poetic.

Practicality, in fact, had enabled him to become an effective war president on the fly. Once the conflict had gotten under way, and once it showed signs of a long and dreadful tenure, Lincoln did his homework. He hunted down military texts on the shelves of the Library of Congress and read late into the night. He sought out generals in order to discuss strategies and tactics. And he got to know guns.

Guns, the president realized, were obviously necessary to the war effort, but his enthusiasm was based on more than that. He also genuinely enjoyed keeping up with the technology of armaments, just as he relished keeping up with other kinds of technology. Lincoln didn't fear or mistrust technology. He employed it in his most crucial duties. Almost every day, the president would leave the White House and head for the side door of the War Department on Seventeenth Street. There, on the second floor, was the telegraphers' room, where messages from battlefield generals were received and transferred onto paper. These reports were handed to an anxious

Lincoln, who would sometimes spend the night in the room right next door, on a couch in the office of War Secretary Edwin Stanton. The telegraph was proving to be a vital link between the president and his generals, and Lincoln wanted to be there when crucial messages came across the wire, broken up into the ticking hiccup of Morse code. Among the obscure telegraphers who handed the president these messages was an ambitious young man named Andrew Carnegie, who recalled that the long-limbed Lincoln, awaiting a reply to some instruction sent to his commanders, had a habit of leaning impatiently against Carnegie's desk. "Intellect shone through his eyes and illuminated his face," Carnegie said years later, "to a degree which I have seldom or never seen in any other."

Lincoln was dazzled by all sorts of gadgets, determined to keep abreast of the latest breakthroughs in not only communication, but also transportation and armaments. He was comfortable with the rapid changes that seemed endemic to the nineteenth century. His early career had been based on just those sorts of transformations. As a young railroad lawyer, Lincoln had helped the nation move from steamboat to rail travel by representing the railroad in court battles with stubborn steamboat owners over bridges and right-of-way issues. As president, Lincoln's eye was always fixed on the technological future; even as the Civil War raged, he announced his support for a transcontinental railroad. He relied on Joseph Henry, who ran the Smithsonian Institution, to keep him up to date on the scientific world. And it was Henry, the self-effacing mechanical wizard, whose electromagnetic relay had turned Samuel Morse's telegraph from a clumsy apparatus that didn't work over long distances into the most effective and reliable communication system the world had yet seen.

Steam locomotives, telegraphs, weapons: They all challenged the president's imagination. He was fascinated by oncoming technological marvels. That is sometimes easy to forget, because the image of Lincoln as the timeless literary man—an image gravely polished by the magnificent public speeches, the graceful and heartfelt private letters—is so powerful, seemingly so untethered to a specific era. But Lincoln was very much a man of the 1800s: America was restless and growing, change was everywhere, and he loved being in the middle of it all.

Why, then, has the popular image of a gun-shy president persisted? Many of Lincoln's biographers still take pains to claim that as an earnest and sensitive young man he disliked guns, that "in a society of hunters, Lincoln did not hunt; where many males shot rifles, Lincoln did not shoot," as if that fact confers a sort of saintly halo atop his craggy head. Lincoln may not have been a casual sportsman, but he was vitally interested in rifles—how they worked, how they were made, what sorts of improvements were on the horizon. And he test-fired guns during his presidency. Lincoln's intense personal involvement in procuring Union armaments is well documented. Perhaps the polarizing stereotypes that have helped to inflame contemporary debates over guns and gun control somehow require Lincoln's admirers to insist, contrary to the evidence, that he was indifferent to firearms.

He wasn't. He was enthralled by them, and he knew a great deal about them. Lincoln's counterpart on the Confederate side, Jefferson Davis, also knew about guns. Davis was an 1828 graduate of West Point. He had served as secretary of war under President Franklin Pierce. He kept up with the latest innovations in armaments, realizing ahead of many of his countrymen that breechloading rifles were superior to muzzle loaders. As war secretary, Davis brought the minié ball to the attention of the ordnance department; that cone-shaped, hollow-based bullet, named for the French military officer Claude E. Minié, was smaller than the barrel diameter, thus easier to load and more accurate because the hollow base expanded when it was fired, enabling the bullet to grip the rifled grooves as it rushed through the barrel. The minié ball would become the signature bullet of the Civil War.

As war secretary in 1855, Davis was just doing his job when he kept up with weapons development, but Lincoln, as president, actually seemed to look forward to visits from firearms manufacturers. Christopher Spencer, inventor of the first repeating rifle suitable for use in combat, came to the White House at least twice. He let a delighted Lincoln test-fire the sleek ten-pound. Spencer carbine. It had a tubular magazine in the butt stock that could hold ten rounds. Lincoln may have been enjoying himself, but the time was well spent; Spencer's rifles would prove to be a crucial factor in the Union victory at Gettysburg. Another gunmaker reported that when he showed up unannounced at the White House, he was "ushered immediately into the reception room, with my repeating rifle in my hand." The surest way to get an audience with Lincoln was to show up with a firearm.

What really intrigued the chief executive was the possibility of a new kind of weapon altogether, a weapon that could fire multiple times in succession. A gun, in other words, that operated the way a machine does: automatically, over and over again. The concept was not new, and a man with Lincoln's abiding fascination with technology would have been well aware of the many attempts to create such a gun. But now his interest was more than academic: Lincoln was convinced that such a tool would be of immeasurable help in the war effort. It could make a critical difference. The president believed in ingenuity, in the power of mechanical advances to change the odds, and he couldn't resist the experimental weapons brought to his doorstep by hopeful inventors. "During the Civil War," noted historian Robert Bruce, "the nearest thing to a research and development agency was the president himself." Lincoln's assistant, William O. Stoddard, kept his office filled with the arms that arrived for the president to fiddle with. The guns were stacked on tables, tilted against walls, heaped onto chairs.

Among the first multiple-firing weapons submitted for Lincoln's perusal was an odd-looking box set between large wheels. The box, from which a thin barrel protruded, was topped by a tapered hopper. It was apparently Lincoln himself who nicknamed the device the "coffee-mill gun," because it resembled the box used to grind the beans. Others called it "the devil's coffee mill" or, less picturesquely, the Union repeating gun. An American inventor of agricultural machines, Wilson Ager, had patented the design in Great Britain, although he did not invent the gun. No one is quite sure who did. Several of Ager's countrymen claimed credit for having invented it, and naturally a lengthy court battle ensued.

Lincoln didn't care who had dreamed it up. He only cared if it worked, and it seemed to work well in the demonstrations that a sales representative named J. D. Mills arranged for the president in the fall of 1861. Mills had a colorful sales pitch; he called the gun "an army in six feet square" or, catchier still, "an army in a box." The gun's steel tubes were fed either with loose powder and balls or with paper cartridges. It sent off some 120 rounds per minute. Lincoln loved the gun. He was amazed and impressed at the intense rate of fire. Overriding the objections of his ordnance department, whose leaders were philosophically and temperamentally opposed to trying newfangled weapons in the middle of a war, the president ordered the purchase of ten Ager coffee-mill guns on October 16, 1861. For the first time in American history, machine guns were part of the nation's arsenal.

Back in Indianapolis, Richard Gatling was still hard at work on his own version of a machine gun. By the time Mills made his initial sale of the Agers, Gatling had been refining his design for more than six months. Doubtless Gatling knew about the Ager gun—the world of inventors was a small one, and the world of arms makers smaller still—yet Gatling was confident that his gun would be superior. The Ager gun and the Gatling gun did share a central feature: Both used steel tubes containing paper cartridges as ammunition for each barrel. But the Gatling gun had a crucial advantage: Because of Richard Gatling's mechanical acumen, his gun would be more efficient and reliable. And because Gatling insisted on overseeing production, his guns were made much more precisely. The trouble was, however, the Ager gun had a head start. Another six months would have to pass, after Mills made the deal to sell the initial Ager guns, before Richard Gatling was even ready to test his gun. Consequently, Gatling would spend most of the war fruitlessly pointing out to ordnance officers—and to Lincoln himself—that the Gatling gun and the Ager gun were very different machines. Test firings at government proving grounds routinely backed up Gatling's claim. But when it came to machine guns and the Civil War, the well was poisoned early: When military men heard "machine gun" they pictured not the steady, solid, dependable Gatling gun, but the clumsy, inept Ager gun. That misplaced prejudice haunts the reputation of the Gatling gun down to the present day.

In the fall of 1861, though, the Ager still looked like a good bet, especially since it carried Lincoln's blessing. After the government's initial ten-gun contract, General George McClellan ordered another fifty for his troops. General Benjamin Butler, the notoriously flamboyant Union officer who later would develop an affection for the Gatling gun, was always on board early when it came to flashy new weapons: He bought two coffee mill guns in 1861, reportedly strapping them on boats for river patrols. The next year, General Frémont purchased two for his unit. For a brief time at the outset of the war, new weapons were popular. The boss liked them—Lincoln spent hours each week at the Washington Navy Yard, reveling in the tests of fledgling firearms—and the innovative weaponry seemed like a fine idea for a war that was sure to be of short duration. Why not try some clever new weapons? What was the harm?

The harm, as the men forced to use the Ager gun discovered, could be substantial. Weapons had to work—not just in the carefully controlled environment of government testing, but in the heat and smash and panic of actual combat. Weapons had to work when lives were at stake, when a jammed or overheated gun or inaccurate firing would be more than simply a temporary inconvenience, when the consequence of failure wasn't just an embarrassed and apologetic J. D. Mills sipping a cup of tea in the White House and explaining all about metal tolerances and barrel diameters, but a dead soldier. Many dead soldiers.

Machine guns weren't a new concept in the 1860s. They were a very old concept. The creation of a weapon that could fire multiple times had challenged inventors over the centuries, enticing the best minds of each era to step up and grapple with the problem, a problem that encompassed many fields, from physics to mechanics to metallurgy to chemistry. A machine gun was the cold fusion of its day: Everybody from crackpots to reputable scientists wanted a chance to come up with it, to take that design leap, to prove the skeptics and naysayers wrong. The quest to make multiple-firing weapons even predated the advent of guns and bullets. In the ancient world, the Greeks experimented with a magazine-fed device to shoot arrows in swift succession. In 1066, multiple-firing bows were employed by the British at the Battle of Hastings. As long as people have faced off on battlefields, better weapons have enticed them. As long as people have been starting wars for all the reasons that they do—religion, property, geography, vengeance, pride, boredom—they have sought ways to kill each other faster and more efficiently.

So the idea of a machine gun was familiar, but for many centuries, nobody knew how to make one that worked. Early attempts were clumsy, flawed, and downright dangerous for the operators, leading to the venerable joke that some guns were deadlier at the breech than at the muzzle. In the fourteenth century, devices called "organ guns" were introduced in European warfare; as many as fifty parallel barrels, laid in a flat row on a wheeled cart, supplied the firepower. Each barrel had to be loaded and lit individually, however, making actual use of the weapon terribly cumbersome and unreliable. Still, the French army dragged along organ guns in 1338 when it attacked the British coast at Southampton. The marauders reportedly brought just three pounds of gunpowder, leading some historians to speculate that the organ guns were more for show, for psychological impact, than for actual use. This idea—using military hardware as a provocative symbol, and winning the day without firing a shot—would be fulfilled as well with Gatling guns in the nineteenth century. The mere sight of lethal-looking weapons could be an effective strategy.

Shortly after the sacking of Southampton, the British rose, dusted themselves off, and began making their own organ guns. Now an arms race was under way in fourteenth-century Europe, with multiple-firing weapons taking up the space that nuclear weapons would occupy in the twentieth century: As long as one king had them, every king wanted them. The era of bigger and bigger and still bigger guns was launched. Pushed along by technical advances in metallurgy that enabled the use of malleable wrought iron in weapons making, the huge, crude cannons known as bombards began to blunder their way onto battlefields, ejecting tremendous stones that could batter down fortifications and create a kind of percussive hell. By the fifteenth century, mammoth weapons were no longer a novelty in warfare. They were a necessity.

Still, the concept of a multiple-firing weapon constituted a special and seemingly intractable problem. If guns were fired too quickly and too continuously, the barrels overheated. And how to create a continuous feed of ammunition, given the slow and ponderous way guns were loaded at the time? A Scottish inventor named William Drummond was granted a patent in 1626 for "a machine in which a number of musket barrels are fastened together in such a manner as to allow one man to take the place of a hundred musketeers in battle," with the muskets arranged in a circle and ignited, one by one. Apparently it was never built. Some four decades later, an otherwise obscure British inventor suggested using the power behind the gun's recoil action, as well as the accumulating gases, to supply the force to load and reload a weapon. This was all quite theoretical, and could not have been built with the technology of the time, but it was the first important employment of physics principles in thinking about how a multiple-firing weapon might work. And then in 1718 came the Puckle gun.

James Puckle's intricate and highly detailed patent application has intrigued firearms historians for centuries. Puckle proposed a gun "that discharges so often and so many bullets, and can be so quickly loaded as renders it next to impossible to carry any ship by boarding." His sketch shows a revolving chamber and a long barrel set atop a tripod, with a hand crank projecting from behind the chamber. Puckle, though, undermined the seriousness of his effort with a proposal that his gun be loaded with round bullets to kill Christians and square bullets to kill non-Christians. His gun, Puckle added in the written material accompanying his design, would be especially useful "in Defending YOURSELVES and PROTESTANT CAUSE."

Advances in weapons technology didn't come in steady, gradual steps. They came in brief, hectic bursts, followed by long static periods. Until the nineteenth century, there had been a lengthy period during which armaments remained relatively unchanged. It was like a spooky, unnerving pause in a battle: Something *had* to happen sooner or later. Didn't it?

It did. In the late eighteenth and early nineteenth centuries, the ground fairly shook from a rapid series of technical developments that would revolutionize much about armaments and clear the way for the first effective machine guns: the change from flintlock to the percussion cap; the rifling of bores, which previously had been smooth, and the switch from muzzle loaders to breechloaders; and the introduction of metal bullets, replacing paper cartridges. There was now a new way to fire bullets; a new place from which to fire them; and a new type of bullet to fire.

With the opening of the Civil War, arms makers had powerful incentive to exploit the new technologies and come up with fresh designs for multiple firing weapons. Along with the Ager and the Gatling, the era would see the production of a number of new machine guns. The Vandenburgh volley gun, invented by a New York State militiaman named Origen Vandenburgh, was a stubby-looking breechloader, its barrels fully encased in a large metallic shell; some versions had a whopping 451 barrels, which could be fired all at once or in smaller groups, creating a ferocious recoil. A disgruntled Vandenburgh sold it to the Confederate Army, but it saw only limited use. Among the problems, as always, was reliability: When the going got tough, early machine guns tended to jam or overheat or both. The Billinghurst-Requa was another odd-looking contraption; its twenty-five barrels were arranged not in a circle but in a straight line, on a flat device that looked a bit like a backgammon board. This design would show up again in 1878, when William Gardner of Toledo, Ohio, patented his Gardner gun; the Gardner ended up being a serious international rival to the Gatling gun in the 1880s and 1890s. However, for the Billinghurst-Requa—nicknamed the "covered-bridge gun"—the future was not bright. Briefly employed by the Union Army, it proved to be fatally flawed on the battlefield, breaking down at regular intervals. (Machine guns seem to attract fanciful nicknames: Early in the next century, a German-made Maxim gun would be dubbed "the devil's paintbrush" and a Browning gun made by the Colt Company would be called the "potato digger.")

Funny nicknames, exaggerated promises, great hopes that soon fizzled out: The world of machine-gun development sometimes has the harmless, slapstick aura of a passel of nutty professors fond of blowing things up in their attic laboratories—until, that is, you remember that the objects being created were intended to maim and kill in greater quantities than ever before.

On March 29, 1862, during a skirmish between Colonel John W. Geary's 28th Pennsylvania Volunteer Regiment and a Confederate squadron, there occurred what is believed to be the first use of a machine gun in combat. Geary and his men had been issued two of the ten coffee-mill guns that Lincoln had insisted on buying. Their assignment was to keep an eye on a stretch of land along the Potomac River near Harper's Ferry. Typically, Geary's unit saw little action, but there was a bit of a tangle with a Confederate band in the otherwise unremarkable town of Middleburg, Virginia, and at some point, the specifics of which were lost in the ensuing dusty cloud of confusion and shouting, a coffee-mill gun let loose and a Southern soldier fell.

Still, the brusque, tradition-minded Geary didn't much like the fancy new guns. A few months later, he would ship them back to Washington, complaining curtly that they were "inefficient, and unsafe to operate." Geary was trusted on the point, which made sense; he had an impeccable record of military judgment. He had been the first mayor of San Francisco and the governor of the Kansas Territory. Duty had sent him far from his native Pennsylvania, including an appointment as the first American commander of the newly conquered City of Mexico. If Geary said that coffee-mill guns weren't worth the trouble, then few in Washington would argue. At war's end, the coffee-mill guns that had once so bedazzled Lincoln were sold as surplus ordnance for a tiny fraction of what the government had paid for them.

The biggest problem with the coffee-mill gun was its workmanship, not its design. These first versions apparently were somewhat shoddily made. They were unwieldy to transport and jammed easily.

They had a maximum firing range of about eight hundred yards, which was not much of an improvement on the ordinary musket. They often overheated. Ager, in fact, would patent a fan-cooled system to keep the barrel from getting too hot—but not until 1866, long after it mattered to Geary and his men.

Still, a machine gun in the field was unprecedented, and some soldiers surely did a double take. What the Ager guns lacked in dependability they made up for in novelty. "I saw one of them work; it was tick, tick, tick, sixty to the minute, as fast as you could think; no brisk little French clock ever beat faster," a soldier on the scene recalled. "When the barrel gets hot, there is another in that chest; when the grists are all out and the battle over, you pack the whole affair in a sort of traveling-trunk, slip in a pair of shafts, with a horse between them, in a twinkling." Yet this eyewitness also noted that for all of its splashy newness, for all its tidiness and compactness, the coffee-mill gun was not beloved by the rank and file: "But soldiers do not fancy it. . . . It is so foreign to the old, familiar action of battle—that sitting behind a steel blinder and turning a crank—that enthusiasm dies out; there is no play to the pulses; it does not seem like soldiers' work."

It does not seem like soldiers' work. It did their killing, but it did not satisfy their souls. This would come to be a crucial point—perhaps *the* crucial point—in the career of the machine gun; indeed, in the coming course of modern warfare. While it is indisputably true that "the machine-gun . . . enables[s] a crew of two or three to produce firepower equal to, if not greater than, that of a Napoleonic battalion," the battalion is human. The machine is not. Another kind of truth would have to be reckoned with: the crushing, dispiriting reality that cold-as-steel machines now might matter as much as or more than hot-blooded men, that human beings would become dispensable, interchangeable, that war was no longer a matter of personal commitment and dauntless courage and gritty endurance but of levers, gears, cartridges, and barrels.

The chasm between what these early machine guns could actually accomplish and how military personnel felt about them was not really a matter of logistics, or efficiency, or official reports from the ordnance department. It was a matter of imagination. It was as much about emotion as about technology—and it would haunt the initial relationship between machine guns and the men who waged war. Those wars had been waged the same way for centuries. A war was a simple, elemental contest between one soldier and another soldier, face to face, eyeball to eyeball. It was basic and bracing. It was easy to understand. Machine guns seemed to interfere with that stark bargain. They complicated things. They robbed the fight of its familiar vigor and dash.

In his imaginative recapitulation of the Civil War, *The Red Badge of Courage,* Stephen Crane very nearly foretold the revolution in armaments that soon would occur—soon, but not yet. "Presently he proceeded again on his forward way," Crane wrote of the young soldier at the heart of his novel. "The battle was like the grinding of an immense and terrible machine to him. Its complexities and powers, its grim processes, fascinated him. He must go close and see it produce corpses." An immense and terrible machine. With corpses as the end products. Crane rightly intuited that machine guns were on their way.

In these early days, though, when the machine guns were still new enough to seem like something that might be successfully resisted, many soldiers were suspicious of them. No matter how well the fancy new guns might work, soldiers and their commanders did not relish their use. Thus the Gatling gun had a fight on its hands before it even left the factory floor.

The coffee-mill guns were a disappointment to Lincoln. They were temperamental and unreliable, and his army, quite naturally, didn't trust them. They were expensive—from $1,300 to $1,500 apiece—and without an enthusiastic reception from men in the field, without a push from the ground up, the exorbitant cost was impossible to justify. So Lincoln let it go. He switched his attention from machine guns to repeating rifles.

That was bad news for Richard Gatling, who had hoped to persuade the president of the merits of yet another machine gun. An even better one. One whose ingenious design and high quality of manufacture made it reliable and accurate. The Gatling gun, as Gatling himself argued and as demonstrations by neutral parties showed, did its job well. "I have seen an inferior arm known as the 'Coffee Mill Gun,' which I am informed has not given satisfaction in practical tests on the battlefield," Gatling wrote in a personal letter to Lincoln. "I assure you my invention is no 'Coffee Mill Gun'—but is an entirely different arm, and is entirely free from the accidents and objections raised against that arm." By this time, though, the nation was hip-deep in the war, and no one was listening. There was no longer an appetite to test-fire and exclaim over shiny new weapons. The passion for experimentation in armaments was gone. This war was turning out to be an epic struggle, a grim and prolonged and expensive catastrophe—not a short, sharp rap on the knuckles of those unruly seceding states. New guns were no one's priority.

The real problem with weapons in the Civil War, moreover, wasn't that there were too few varieties. There were too many. On both sides, it was a veritable carnival of arms—some new, some old, some muzzle loading, some breechloading, some up to date, but most a motley assortment of whatever could be grabbed in a pinch. Many soldiers brought their own guns. Others scavenged any firearms they could find. Between August 1863 and June 1864, soldiers in the Army of Tennessee clutched smoothbore rifles, .58 Springfield rifles, .70 Belgian rifles, Spencer repeaters, .52 Hall rifles, and many others. Union soldiers used more than twenty-five different types of cartridges.

There was a great blaze of different guns, guns of different shapes and sizes and colors and ranges, and a constantly shifting bazaar of different ammunition; there was the old and the new, the traditional and the innovative. The Civil War constituted a great crossroads in armaments history. A Northern unit at Gettysburg, notes historian Paddy Griffith, sported smoothbore rifles, muzzle loaders, and breechloaders, which means there were "almost two centuries of small arms development concentrated in a single regiment."

In the first half of the 1800s, the flintlock had given way to the percussion cap; the latter was a definite improvement over the firing mechanism found in the muskets of the Napoleonic era. Yet the new technique of loading and firing still was iffy and perilous. The percussion cap pistol required the shooter to fit the cap—filled with fulminate of mercury—on a nipple connected to each of the gun's chambers. Inside the chamber were lodged gunpowder and a lead ball, awaiting the hammer to hit the cap and set off the explosion inside the chamber. These pistols were so tricky to prepare for firing that Jesse James, who was intimately familiar with their workings, still reportedly lost the tip of his left middle finger from a loading mishap. Muzzle-loading rifles were even worse. A Civil War soldier clasping a muzzle loader in the middle of fierce fighting, with bullets whisking past his ears like darting squads of lethal mosquitoes, with friends falling in ragged heaps at his feet, amid smoke and peril, had to retrieve a paper cartridge with the powder charge and bullet; tear it open; pour the powder into the barrel; stick the ball point into the bore; push the ball down the muzzle with his ramrod; remove the old percussion cap and replace it with a new one; cock and fire. A man who had performed the frustrating and tedious procedure many, many times could possibly fire—if he were lucky, and if he were not too unnerved and distracted at being in the middle of frantic combat—three rounds a minute. And firing these guns carried a high physical cost, in addition to the danger. The recoil left shooters' shoulders purple and swollen with bruises. Faces were stained deep blue from the exploding powder. After an especially long battle, a Confederate soldier recalled that his "arm was all battered and bruised and bloodshot from . . . wrist to shoulder, and as sore as a blister." The gun, he added, "became so hot that frequently the powder would flash before I could ram home the ball, and I had frequently to exchange my gun for that of a dead colleague."

Couldn't people see that loading a gun from the back would be better than loading it from the front? Of course they could. In fact, the first real breechloading rifle in the United States had been patented way back in 1848, by Christian Sharps. But firearms development wasn't just a matter of figuring out what worked. It was a matter of figuring out what worked and then figuring out how that improvement could be replicated on a large scale, without losing quality, and at reasonable cost. A breechloader was a far more complex instrument than a muzzle loader. Machine tooling and precision techniques were required to make a breechloader, in order to seal the breech to prevent gas from escaping. The same dilemma would occur with machine guns: manufacturing techniques had to catch up with design. Dreaming up a fine new multiple-firing weapon was one thing; but transferring the idea from a sleek, perfect drawing to a tangible object, and then being able to duplicate the making of that object again and again and again—this was a *gun,* after all, not a one-of-a-kind piece of art—was daunting. And there was yet another impediment to the widespread use of breechloaders during the Civil War. Such guns were, perversely, victims of their own success. Many commanders were frankly afraid that breechloaders would give soldiers an excuse to be sloppy marksmen. Because the men knew they could fire many times in succession, the officers reasoned, they would not take the time to aim with care.

Similar economic and logistical challenges came into play with rifling bores. It was one thing to know that a rifled bore enabled the bullet to fly a greater distance, with more accuracy. But retrofitting thousands of smoothbore rifles with rifled bores—created by cutting helical grooves on the inside of the barrel, grooves that force the bullet to spin along its axis, parallel to the barrel—was a huge and expensive undertaking. Anyone could see that a rifled bore was superior, a superiority that arises from a physics principle known as the conservation of angular momentum: The spinning motion of the bullet prevents the bullet from being deflected by the air resistance surrounding it. But possessing the knowledge, and being able to apply that knowledge to weapons development in the middle of a howlingly ferocious war, is something else altogether. Both sides had begun the Civil War terribly short of arms. There was no time for visionary, chin-stroking thoughts about better weapons; this was a crisis, and in a crisis, you make do.

In the meantime, men fought with the weapons they had, and the weapons they had were often broken and dirty and jammed, and even when they were in the best condition, still required an enormous amount of time to load and fire. A Civil War battlefield was different in many ways from the battlefields of subsequent wars, and one of those ways was aurally. Intense shooting could not have been sustained for extended periods of time. Guns took so long to reload that the frequency of shots would have sounded, to modern ears, almost comically low. Each soldier carried some forty rounds. The rate of fire could have reached one round per minute, but if that were maintained, a regiment would have been out of ammunition in less than an hour. Several battles raged for many hours, even days. That means that most soldiers shot only once every five or ten minutes, to save bullets. When they ran out of ammunition, they resorted to the bayonet charge, or even to rock-throwing; Confederate units were known to do just that. Consequently, instead of sounding like a continuous racket of attacking fire, a steady and terrible din, a Civil

War battlefield actually would have sounded more like a discordant symphony of sporadic pops. That is just how soldiers described the uneven clamor of a firefight: "pop-popping." The pauses, the spaces between the bullets, must have seemed achingly long, at least when they didn't seem terrifyingly short. The only predictable aspect of the noise would have been the unpredictability. And if artillery had been drawn into the fight as well, the intervals between the firing of the shells—necessitated by cumbersome loading procedures—also would have been fraught with a sort of communal held breath, as everyone *waited . . . waited . . . waited* for the next tremendous hit, the gut-jiggling and mind-emptying reverberation. There was an eternity embedded in the spaces between the bullets, a hesitation that marked the precise line between life and death for a random soldier, a soldier who might be standing there one minute, filled with fear and hope and memories of home, and then the next minute, when the shooting resumed, not. It was that fast, and that slow.

A machine gun threatened to change not only the way the battles were fought, but the way the battles sounded. It would change the essential rhythm of warfare, because the noise of its firing was uniform and relentless. Conflicts would no longer sound the way they had sounded, going all the way to the age of the flintlock, when a soldier who achieved three or four shots per minute was considered a nimble prodigy. In the Peninsular wars fought by Great Britain in the early 1800s, the British soldiers "raised a shout of delight" when a bullet hit home; so slow was the pace of the firing that there was ample time for cheers and back-slapping between discharges of rounds. But a machine gun, with its perpetual barrage, with its hailstorm of not only bullets, but the sounds of those bullets, linked and unrelenting, would utterly change the way the battles were experienced by the men locked inside them. And while the Civil War clashes may have been bloody and terrifying, at least they were bloody and terrifying in familiar ways. "I cannot give you particulars or write more now," was a line in the letter that a Wisconsin soldier named Frank Haskell sent to his brother in August 1862 after a fearsome contest. "The terrible weariness of a long fight is upon me." He didn't really have to describe it. Previous letters had done that. Each battle was the same, and there was a consolation in the sameness. As horrific as it all was, it was a horror that was well known, and the pace and cadence of the fury were predictable. It was hell, but it was their hell. A machine gun—that strange and unwieldy interloper—erased the spaces between the bullets. "The rattle of machine-guns," Erich Maria Remarque would write of World War I combat, "becomes an unbroken chain."

The Army didn't want the Gatling gun. And there was virtually no chance that it ever would—at least, not as long as the officer in charge of procuring weaponry was Lieutenant Colonel James W. Ripley, a man who was grimly, vigorously, unalterably opposed to the slick new contraptions that had so enraptured Lincoln. Ripley didn't care how besotted the president was with machine guns. He didn't care how many ordnance officers brought back favorable test reports. Ripley was unmoved. He wouldn't consider them. Between November 1862 and June 1863, ambitious inventors patented more than eighty designs for new machine guns; military personnel tested only seven, and those tests occurred either on the sly or in the face of Ripley's vehement protests. This stern, sharp-eyed soldier with a steep forehead, high cheekbones, and a keen knife-blade of a nose looked like the human equivalent of the word "no": no to new weapons, no to experimentation, no to anything but what the army had always done, in just the way it had always done it. Ripley was notoriously resistant to any sort of innovation in firearms. The mere mention of it could provoke a boiling rage. One visitor to his office described him as possessing "very white hair and a very red face."

The sixty-six-year-old Ripley had been appointed chief of ordnance the same month that the Confederates opened fire on Fort Sumter. He would keep that position until 1863, when he was replaced by Brigadier General Alexander B. Dyer, who stayed for the next decade and proved to be far more of a visionary about new armaments. But it was Ripley who made the initial decisions about machine guns in the Civil War. It was Ripley who staunchly defended the status quo. He set the tone early, with his June 11, 1861, memo, in which he referred to experimental arms as "a great evil," adding that the profusion of new weapons had been "producing confusion in the manufacture, the issue, and the use of ammunition, and [been] very injurious to the efficiency of troops. This evil can only be stopped by positively refusing to answer any requisitions for or propositions to sell new and untried arms, and steadily adhering to the rule of uniformity of arms for all troops of the same kind, such as cavalry, artillery, infantry." Military men looked to the past, not the future, for guidance and inspiration. Prior to World War I, "The sense that victory in war depended on technological progress," notes historian Martin Van Creveld, "only slowly made its way up the military-political hierarchy." Tradition and precedent had the whip hand. Or as Barbara Tuchman put it, "Dead battles, like dead generals, hold the military mind in their dead grip."

In one important way, Ripley did display a bit of forward thinking. He required that all small arms used by government troops be made with interchangeable parts. Such was the key to the American system of mass production that was just taking hold, soon setting the standard for manufacture throughout the rest of the world. And for all of his moody, quick-tempered, authoritarian brusqueness, those who knew Ripley best also knew that he carried the weight of constant sorrow. Only three of his nine children lived to adulthood. When his four-year-old son Roland died in 1838, Ripley quietly folded the boy's clothes, placed them in a small trunk, and carried the trunk with him from military posting to military posting, from city to city, for the rest of his life. Ripley was a soldier; he had a soldier's impassive bearing and stubborn refusal to yield. He did what he did because he thought it was the right thing to do. And machine guns, he believed, were gimmicks, not serious arms. They were distracting dalliances, hardly worthy of serious attention. Especially not in wartime.

When Ripley posed iron-jawed and stiff-backed and unblinking in the middle of the road, blocking the advent of new weapons, he was really just doing what military personnel have always done: Resisting change. Holding fast. Standing firm. There is an inherent conservatism to military thinking; that mind-set is reflected in the old adage that generals are always perfectly prepared to fight the last war. "The army," charged military historian David A. Armstrong, "lacked the bureaucratic, fiscal and above all the intelligent capacity needed to recognize, explore and exploit the potential of novel weapons. . . . Military leaders seldom attempted to accelerate the creation of new arms and concepts for their use."

But Gatling had to try. And to an inventor with a new kind of gun and hopes for a large government order, Ripley was the gatekeeper. Ripley was the key. So Gatling wrote a courteous letter to the ordnance department. He asked that his guns be tested. The reply: a cold, curt, unequivocal "no." Gatling, undaunted, moved on. His next step was to contact generals in the field. Just take a look, he said. Just put the Gatling gun through its paces. You won't be disappointed.

Gatling's persistence was completely in character. It was the way he had always been: indomitable, energetic, a sunrise made flesh. It was as normal for an R. J. Gatling to be gamely optimistic as it was for a J. W. Ripley to be stubbornly obtuse. Gatling had too much pride and capital riding on his gun, too much faith in his abilities as an inventor, to quit. He had already endured setbacks that might have sent a less confident fellow into another line of work. After all, the first half dozen Gatling guns ever made had been destroyed by fire in Miles Greenwood's foundry back in Cincinnati, a total loss. Gatling had bounced back from that early blow, contracting with the Cincinnati Type Foundry to manufacture thirteen new Gatling guns from scratch.

And then: a thread of light beneath the closed door. For a brief moment, Gatling's perseverance seemed to be paying off. In 1863, those thirteen new Gatling guns were sold. They were sold not to Ripley's ordnance department, but to two Union officers. Major General Benjamin F. Butler—that swashbuckling, larger-than-life scallywag who always kept an eye peeled for the latest gadget, the brightest new weapon, and who had also tried the Ager gun—snapped up a dozen, and Admiral David Dixon Porter bought one as well. Porter planned to affix it to the deck of a gunboat, to try for quick victories in skirmishes along the rivers.

The early sales to Butler and Porter might have encouraged Gatling to believe he had broken through the military's steadfast wall of resistance. He hadn't. There was only a flutter of other sales until after the war. No large-scale order from the ordnance department was forthcoming. Yet Gatling, ever the optimist, kept pressing and pushing and then pushing harder. He continued to call upon his influential friends for testimonials. He demonstrated his guns wherever he could. He wrote letters and took out ads and made public appearances.

Having been rejected and rebuffed at every turn—while field tests continued to prove that Gatling guns were efficient and accurate, that they worked smoothly and well, that they were clearly superior to all other kinds of machine guns—Gatling went right to the top. On February 8, 1864, he wrote to Lincoln himself: "His Excellency, A. Lincoln, President of U.S." There was a quietly desperate edge to this letter. There was a lingering sense of righteous indignation that he could not get a fair trial for his gun, a decent hearing. On display in the letter were all the attributes that seemed to coexist in equal measure in Richard Jordan Gatling: bombast, humility, reason, resolution. Here was the inventor, the salesman, the patriot, the entrepreneur. "The arm in question, is an invention of no ordinary character," Gatling wrote. "It is regarded, by all who have seen it operate, as the most effective implement of warfare ever invented during the war, and it is just the thing needed to aid in crushing the present rebellion. The gun is very simple in its construction, strong and durable, and can be used effectively by men of ordinary intelligence." Gatling finished his pitch: "May I ask your kindly aid and assistance in getting this gun in use? I know of a truth that it will do good & effective service. Such an invention, at a time like the present, seems to be providential, to be used in crushing the present rebellion." And just in case Lincoln was confused, just in case he was under the misapprehension that one machine gun was very much like another, Gatling's addendum to his letter, quoted above, gently but firmly differentiated between the coffee-mill gun and the Gatling gun.

There is no record of a reply from Lincoln. The war was going badly, and the president had other concerns. Machine guns were rather low on his priority list right then, especially since the Ager guns had been such a bust. Especially since Gatling guns were often being bad-mouthed by tradition-minded ordnance officers. Gatling guns, in fact, were gossiped about like debutantes; there was a touch of spiteful malice in the whispers. Gatling guns were the focus of innuendoes, of nasty rumors and ill-informed speculation. It wasn't fair, but it worked.

Benjamin F. Butler didn't care about the gossip. He didn't care what anybody else thought about the Gatling gun. Actually, he didn't care what anybody else thought about much of anything. He was a blunt, stubborn man, willful and eccentric, a canny operator, a character, a flashy rascal. As a Union general in the Civil War, Butler parlayed his political connections into appointments to a series of important commands, most of which he failed at, and failed in uniquely colorful ways. He was rude and headstrong but unflappable. Appointed military governor of New Orleans after Union forces took control of that city in May of 1862, Butler proved to be spectacularly corrupt and grotesquely tyrannical; he swiped gold bullion, hanged a man for hauling down an American flag, announced that Southern females who ridiculed his men would be treated as prostitutes. By December, Butler had been sent packing, transferred to another post. Yet for all that, Butler was the only general to purchase and use Gatling guns in Civil War combat. He was, in this instance, a visionary.

From June 15, 1864, to March 25, 1865, Union and Confederate troops were locked in a deadly stalemate outside Petersburg, Virginia, a city south of Richmond, the Confederate capital. Robert E. Lee knew he had to hold Petersburg. Five railroads converged there, making it a major supply route, as well as the last line of defense for Richmond. Ulysses S. Grant knew that, too. Month after month after month, it went on. Northern troops pushed, and Southern troops resisted. Casualties were staggering: some 42,000 for the North and 28,000 among the Confederates. It would be dubbed the siege of Petersburg, and it would come to symbolize the whole mad enterprise of the Civil War: surges, the repulsions of those surges; another surge, another repulsion. The gain of a few precious inches of ground. And then the loss of those inches, and maybe a few more, while the bodies jumped, twitched, and fell when the bullets struck them.

General Butler was in charge of some 30,000 Union soldiers along the James River, northeast of Petersburg, with personal orders from Grant to sweep into Richmond. Butler was equipped with a dozen Gatling guns, which he had purchased for a thousand dollars apiece with his own money, after the ordnance department had turned down his request for funds. "I saw General Butler testing the Gatling Gun," recalled a Union officer, "which was a new thing then, upon some unarmed and unsuspecting Rebels who were strolling up and down the top of their earthworks talking to our men in the rifle pits. It brought on quite an artillery duel." But it didn't make anyone in the ordnance department sit up and take notice, or change any minds. It felt like a fluke, like a diversion from the serious business at hand.

Indeed, the only place in which a Gatling gun was destined to make an appreciable difference during the Civil War wasn't on a battlefield at all. It didn't come in the midst of a struggle between competing armies. And it would set the tone for the weapon's dark reputation later in the century, for its grim identification with the forces of oppression and exploitation.

"City is in intense excitement. Business all suspended. Rioting in almost every ward of the city." That was the urgent message Colonel Robert Nugent telegraphed to a military official on July 14, 1863. "It is a Spontaneous movement. There Seems to be no organization." It was here, in this sideline crisis of the Civil War, far from the bloodstained acres of Gettysburg or Shiloh or Chancellorsville, the sites of so much nobility and heroism, that the Gatling gun played its most decisive and memorable role. It was surely not the glorious battle Gatling had envisioned for the triumphal public staging of his gun.

New York City was in chaos. For three days the tension had risen ominously, from Saturday, July 11, to the Tuesday of Nugent's telegram. Finally it exploded in rioting, arson, looting, and the utter unraveling of the nation's largest and most prosperous city. The cause of the mayhem: the military draft. Five months earlier, Lincoln had signed the Enrollment Act, which threatened to scoop up every man between the ages of twenty and forty-five—unless he paid a three-hundred-dollar fee or hired someone to take his place.

To the poor, to those who would never in their lives be able to get their hands on three hundred dollars all at one time, the unfairness of the draft felt like a blindsiding blow. It felt like a grievous insult. Was there one law for the wealthy and another for the rest of America? Tycoons such as J. P. Morgan, who paid the fee without hesitation, became known derisively as "$300 men." Other privileged individuals eligible for the draft who bought their way out of the war were Andrew Carnegie, John D. Rockefeller, Jay Gould, Collis P. Huntington, and Jay Cooke.

When officials gathered on that Saturday morning in New York's Ninth District to hold the draft lottery, the fuse was lit. Crowds, muttering darkly, formed on street corners. Weapons were gathered. The city was a brooding stew of resentment. People began "tearing up rails, cutting down telegraph poles, & setting fire to buildings," one eyewitness wrote. Military troops that normally would have been on hand to keep order had been sent to Gettysburg; they had yet to return from that terrible battle in the small Pennsylvania town. An aghast and overwhelmed New York police force was all that stood between the rioters and the destruction of the city. As the violence escalated, the authorities fired a cannon at the crowd, killing at least twenty-seven people and injuring many hundreds more.

On July 17, army units finally arrived to restore order. But not before the city suffered more than two million dollars in damages. And not before there was one last confrontation—a standoff that demonstrated the deeply symbolic power of the Gatling gun, even in these early days.

The mob had lurched and churned toward the *New York Times* building, chanting, jeering, ready to set it aflame. They were incensed at *Times* editor Henry Jarvis Raymond, staunch Republican, friend of Lincoln, proud author of strongly worded editorials decrying the rioters' actions. Informed of the crowd's approach, Raymond set up Gatling guns—two at the north windows, one on the roof. The rest of the *Times* staff was armed with rifles. "Give them grape [shot], and plenty of it," Raymond growled.

When the livid citizens approached the newspaper office, their gaze rose up and up and up to the rooftop, where they saw the gleaming barrels of the ultimate deterrent: a Gatling gun. Reportedly, Raymond himself manned one of the guns. The mob backed down. The story foretells the way Gatling guns would be deployed after the war: as menacing symbols, as icons of sheer destructive ferocity, even if they just sat there.

The war, at long last, was winding down. After nearly ten months of pounding by Grant and his vastly superior forces, the Rebels finally had abandoned Petersburg, Virginia, their last stronghold, at the end of March 1865. Shortly thereafter came the quick and ugly evacuation of Richmond, the Confederate capital. President Jefferson Davis and the rest of his government clawed their way through confused and distraught crowds, taking the last few seats on departing trains; Confederate soldiers did what soldiers fleeing their own base are supposed to do, which is to burn supplies so that the enemy might not confiscate them. On the night of April 2, flames reached into the black sky as offices, warehouses, bridges, and homes were set alight. Terrific bangs, pops, and vibrations signaled that the fire had reached the storehouses of ammunition.

On April 4, a Tuesday, the ruined city had a visitor: Abraham Lincoln. The man whose ship—the U.S.S. *Malvern*—led the president's vessel, the *River Queen,* to the Richmond riverbank was Admiral David D. Porter, only the second person in military history to purchase a Gatling gun. This trip, however, had nothing to do with weapons; the muted, somber Lincoln was touring a defeated city, hoping to speed up the Confederate surrender.

And there was no reason, in any case, for anyone to be thinking about Gatling guns on this day. Gatling guns had been used here and there during the siege of Petersburg, the last drawn-out Rebel stand, but sporadically, almost whimsically. The men under General Butler—whose purchase of Gatling guns had come just ahead of Porter's—would fire them when Butler ordered them to, and apparently a few Gatling guns were discharged during the long stalemate, but without decisive effect. The artillerymen did not really understand how to operate them. The guns looked odd and unwieldy. And so the soldiers' hearts were not in it—a fact that, in the ensuing decades, actually would come to matter a great deal to the destiny of the Gatling gun.

Because as both Lincoln and Lee had come to know so well in the prolonged course of this war, a soldier's heart was vital. His heart was perhaps the most essential part of him—more crucial, even, than foot or knee or shoulder or even trigger finger—because without the heart, without the passion, he will not fight. And with it, he will fight long past the point when it makes logical sense to do so. Many of Lee's troops had kept on fighting against absurd odds and in terrible conditions; they had kept on fighting even after the outcome was inevitable, even as they marched on bloody bare feet and clutched sticks and rocks as weapons of last resort. They did that because of their passion. Now, though, with the collapse of Petersburg and the Confederate government's unseemly flight from Richmond, the end was here.

Lincoln's walk through the stricken city would become famous, one of those moments that unfurls long past its actual length—about a mile and a half—and stretches into the golden infinity of legend and myth. The streets were lined with African Americans shouting, "Glory! Glory! Glory!" There was hand shaking and hem touching and a kind of murmuring ecstatic wonderment that, yes—*yes*, at last—it was over, it was truly over, and here was the man who had changed the nation forever.

Lincoln's route led him to the Confederate White House, and he sat in Jefferson Davis's chair and received the Confederate representative—yet the latter seemed coy and evasive on the question of surrender, especially when it came to the issue of recognizing emancipation. An impatient Lincoln put it in writing: "If there be any who are ready for those indispensable terms, on any conditions whatsoever, let them say so and state their conditions, so that such conditions can be distinctly known and considered." Lincoln, like so many of his countrymen, was sick of the war and desperate for it to end, and in these last throes, these dwindling days, with their exasperating hints of delay and the implied need for further negotiation, Lincoln must have felt a twinge of dread, his own private wonderment: *Was* it over?

Yet the end, when it came, was every bit the swift, simple event that a numb and weary and grieving nation so desired. Five days after the president's walk through the streets of Richmond, Generals Lee and Grant met for two and a half hours on Sunday, April 9, 1865, at the home of Wilmer McLean at Appomattox Court House. Lee arrived first and waited in the front parlor. After Grant joined him, the two chatted privately for a short time. When their respective staffs were motioned in, several observers were struck, as Grant himself initially had been, by the vivid contrast between the two leaders: Lee, silver-haired, stiff, spotless of uniform and shiny of boot; Grant, brown-haired, shambling, his boots spattered and crusted with dried mud. Documents were prepared. Then it was over.

There was, however, one more telling contrast in physical appearance: From Lee's waist dangled an elaborate sword with a jeweled hilt. It was the sight of that sword, Grant later intimated, and all that it represented—honor and dignity, the very essence of the soldier's creed—that helped persuade him to let the Confederates avoid the "unnecessary humiliation" of having to hand over "the side-arms of the officers, nor their private horses or baggage."

In his order to that effect, Grant added: "Each officer and man will be allowed to return to his home." It was a sword—that most ancient and least innovative of weapons, a sword that did not leave its scabbard that day—that did what no gun would have been able to do. It was a sword that helped persuade Grant to let Lee and his men go home without punishment. How people *felt* about weapons, it seemed, could matter nearly as much as what those weapons did, or could do. No matter how efficient a weapon might be, when it came to a soldier's embrace, efficiency was not the only measure.

Even in war—supposedly the most unsentimental of arenas, supposedly a realm filled exclusively with hard realities—symbolism, soft-edged, intangible, still was significant. It was a lesson that Richard Gatling was to learn all too well in the second half of this momentous century, as he set about marketing the invention that carried his name and sealed his fate.

CHAPTER SIX

"A Little Gatling Music"

I remember the ceaseless bombing thunder that shook the house, like an earthquake, the futile popping of revolvers, the whining shells overhead . . . the heavy breathing of my men about me, and always just in front of us, the breathless whir of the gatling.

—Richard Harding Davis, *Captain Macklin*

On March 31, 1887, a Thursday, an unusually exotic and decidedly eclectic collection of items was systematically loaded onto a transatlantic steamship known as the *State of Nebraska.* It was a memorable manifest. There were "97 Indians, 180 horses, 18 buffalo, 10 elk, 5 Texan steers, 4 donkeys, and 2 deer." There were rifles and fringed buckskin jackets and colorful neckerchiefs and big-brimmed cowboy hats. And there were a few other things, too: Gatling guns.

The objects constituted the packing list for Buffalo Bill's Wild West and Congress of Rough Riders of the World, a traveling show that was taking advantage of its biggest break yet: a chance to perform in London as part of Queen Victoria's Jubilee. While a thirty-six-piece band on deck—the musicians tricked out in cowboys hats and moccasins—struck up "The Girl I Left Behind Me," the steamship fumed and churned out of New York harbor.

In the roughly two decades that had passed since the end of the Civil War, the Gatling gun had become a ubiquitous symbol of

American pluck and bluster, of American determination to tame the rugged West, and it was a crucial part of William Cody's show. Since the early 1870s, Cody had entertained thousands of his countrymen in towns large and small with the bombastic extravaganza, complete with sharp-shooting heroes, thundering wagons, whooping Indians, and an electric atmosphere. But this trip was something else again: This was royalty. This was a publicity bonanza. This was the threshold of a global marketplace, the first step in mounting the sort of large-scale spectacle that could bring great profits. Cody's show even had the interchangeable parts of which American manufacturers were always boasting: horses, Indians, cowboys, Gatling guns. If one dropped out, you simply inserted another in its place. There were irreplaceable stars such as Annie Oakley, of course, and there was Cody himself—vivid, flamboyant, unforgettable—but the heart of the show was its clockwork modularity: so many Indians (*check*), so many cowboys (*check*), so many Gatling guns (*check*).

Mounted atop a wagon, barrels flashing as a cowboy fired blank rounds, the hand crank going round and round in the energetic revolution that, to the delight of the audience, would under real-life circumstances have spelled swift and certain death, the Gatling gun was a guaranteed showstopper in the climactic scene in which the brave settlers were saved from the bloodthirsty Indians. Cody knew what audiences liked. He knew what people came to see. He knew how much they enjoyed being frightened and awed. And nothing looked meaner, scarier, and yet more alluring than a Gatling gun.

Until the Thompson submachine gun in the 1920s, and the AK-47 later in the twentieth century, no other firearm could rival the iconic status and instant familiarity of a Gatling gun in the late nineteenth century. It showed up in poems and in novels, in paintings, in newspaper columns, and it made constant cameo appearances in anecdotes and tall tales. Because Richard Gatling understood the value of publicity and never missed a chance to promote his invention, and because the gun itself had simply captured the public fancy in that mysterious, ineffable way that happens with some products and not with others, the Gatling gun was more than a *thing:* It was a catchphrase, a brand name that meant blood and thunder. It represented a don't-mess-with-me mind-set that appealed inordinately to Americans. Thus whatever disappointment Gatling had felt at the government's failure to embrace the gun for Civil War service had surely dissipated by now, banished by the gun's subsequent success both commercially and culturally. Dozens of countries around the world were lining up to buy it. When you said "Gatling gun" these days, people knew what you meant—and with a delighted shiver, they might picture the sinister sweep of those multiple barrels. A Gatling gun meant bold firepower for what was then thought to be a good cause: blunt justice in the wide-open West.

Yet alongside that frontier image, the one that evoked clear-eyed, clean-limbed cowboys and their square dealing, another image had developed in reference to Gatling's invention, a less savory one; in fact, it was a downright ugly one. It had begun on the streets of New York during the draft riots of 1863, when *New York Times* editors reportedly pointed Gatling guns at the marchers, and it continued through the 1870s, 1880s, and 1890s as Gatling guns were purchased by police departments, state militias, and factory owners. Gatling guns now were seen as tools of domination and intimidation, both at home and abroad. During the Pullman strike of 1894, the editors of the *New York Times* referred to a court injunction severely limiting strikers' rights as "a Gatling gun on paper." There was a clear that'll-teach-'em! ring to the editorial, the smugness of money and power. And when the British army announced in 1873 that it would go after the Ashanti people of Africa, the *London Times* urged soldiers to "treat them to a little Gatling music." The Gatling gun, in good ways and bad, was a household name now. A great many people knew what it was and what it could do.

So it was no wonder that Cody made the Gatling gun part of his act. Just after the steamship bumped up against the Albert Dock at Gravesend on April 14, 1887, Cody's staff unloaded the paraphernalia—animate and inanimate—that had been carefully packed at the other end: buffalo, elk, Indians, deer, donkeys, Gatling guns. This time, the band played "Yankee Doodle." Three trains then hauled the entourage to the Midland station, close to the twenty-three-acre site of the American Exhibition. This perfection of logistics was one of the secrets, workaday but crucial, of Cody's great success. He knew how to load and move his massive operation quickly and efficiently from place to place, be it from one American city to another or across oceans and continents. Every detail was attended to. Everything was planned and charted, Annie Oakley would write later, including "every man's position, how long it took, how we boarded the trains and packed the horses and broke camp; every rope and bundle and kit."

The show was a hit even before it opened. In the weeks between Cody's arrival and the first official performance, while his employees set up everything with down-to-the-inch precision, British aristocrats arrived at the site at regular intervals to see what all the fuss was about. Most were agog. Lady Randolph Churchill, actress Ellen Terry, members of Parliament, dukes and duchesses: All came, all raved. On April 25, former Prime Minister William Gladstone made a wide-eyed tour. For years afterward he would tell the tale, with enormous relish, of having shaken hands with a genuine Sioux Indian chief, Red Shirt. (When the Wild West show moved on to the Paris Universal Exposition in the spring of 1889, French couples vied for a chance at the same experience; confident rumors had spread that touching an Indian would instantly boost fertility.) On the actual day of the former prime minister's flesh-pressing moment with Red Shirt, Gladstone was inspired to offer a lunchtime toast to the bond between Great Britain and the United States. Days later Albert Edward, the Prince of Wales and future king, along with his wife and daughters, came by and watched a rehearsal. Annie Oakley was summoned to meet the royal family; the young Princess of Wales, Oakley informed the reporters who insisted on her impressions, was "a wonderful little girl."

On May 11, 1887, Queen Victoria herself attended a performance at the Earl's Court arena. She had been encouraged to do so by her son, the Prince of Wales. It was a headline-snaring day, because the queen had not been spotted at a public event since the death of her beloved husband a quarter of a century before. Cody's show, complete with the deadly-looking but still irresistible Gatling guns, was thrilling enough to draw royalty out of mourning.

It was also thrilling enough, Cody would claim, to affect international relations. The show got under way with a lone rider on horseback galloping dramatically across the arena, brandishing an American flag. That brought the queen lumbering up out of her seat for a quick bow. The obedient crowd naturally did exactly the same thing, while British military officials saluted—and more than a century of simmering discord between the two nations instantly ended, or so the opportunistic Cody would declare: "All present were constrained to feel that here was an outward and visible sign of the extinction of that mutual prejudice. . . . We felt that the hatchet was buried at last and that the Wild West had been at the funeral." Cody may have been overstating things, but not by much. And thus it was an extravagant spectacle—not a treaty signing, not the resolution of a border dispute, not a trade agreement—that erased boundaries in the late nineteenth century. It was an elaborate performance that permanently sealed the American West in myth, in a highly ritualized series of violent symbols: fleeing settlers, nasty Indians, repeating rifles, and Gatling guns. Differences of language and culture and custom evaporated amid the noise and frenzy of the Wild West. At a second show—mounted at the request of the queen—the European royalty that had assembled for her Jubilee was equally enthralled, from the kings of Belgium, Greece, and Denmark to William II, future ruler of Germany. During the big finish, the kings, it was gleefully reported, joined Cody on the stagecoach, with Indians in mock pursuit.

More than a million spectators took in the show before its closing performance in October 1887. Cody, though, was just getting started. He carted it to more cities in England—Manchester and Birmingham—for sold-out performances and then headed across the continent. In France and Spain and Italy and Germany, audiences flocked, cheered, and returned again and again for more. One early fan was Mark Twain, who had witnessed Cody's show in the United States and just ate up the idea that other cultures were getting a taste of America through the Wild West: "It is often said on the other side of the water that none of the exhibitions which we send to England are purely and distinctively American," he wrote. "If you take the Wild West over there you can remove the reproach."

Cody knew what he was selling, and it wasn't just tickets and souvenirs. It wasn't just spectacle, melodrama, and excitement. It wasn't just cowboys and Indians. It was the core image of America, a land whose inhabitants were tough when they needed to be, and always heroic. Always on the side of right. "The bullet," ran some salient sentences in an official program for the Wild West show, "is a kind of pioneer of civilization. Although its mission is often deadly, it is useful and necessary. Without the bullet, America would not be a great, free, united and powerful country."

Cody realized early on how weapons such as rifles and Gatling guns, combined with zesty advertising aimed at the provocative new idea of a single mass market, could make his fortune. "Cody," the wily entrepreneur cackled to himself while watching the London crowds go wild, "you have fetched 'em."

Like Cody, Richard Gatling knew a thing or two about marketing, about the need to never let up. To push, push, *push.* To make, to sell, to keep a constant eye on the competition. Gatling had gotten rich from his gun, but still he never stopped. He redesigned the Gatling gun dozens of times throughout the 1860s and 1870s and 1880s and 1890s, adding and subtracting and refining elements. In 1883, when he was sixty-five years old, he would come up with a version of the Gatling gun powered by compressed air; in 1890, he would replace the hand crank with an electric trigger. In between the major changes were the minor ones, hundreds of them, small and trifling to everyone but him; he never abandoned the urgent search to make the Gatling gun work smoother and faster and more reliably. Getting rich was nice, but it was a side benefit. Getting rich seemed to be only marginally related to who Richard Gatling was or what he really wanted. He was an inventor, a creator.

It had pleased him, no doubt, when the U.S. Army finally came around, officially adopting his gun in 1866. New men ran the ordnance department by then, younger men, men who weren't threatened by innovations in arms. It surely must have been gratifying for Gatling to read official government reports such as this one, contained in an 1874 bill to appropriate $292,600 to buy 209 Gatling guns: "After thorough and exhausting trials at Fort Monroe, Va.," the gun was distinguished by "its lightness of its parts; the simplicity and strength of its mechanism; the rapidity and continuity of its fire without sensible recoil; . . . its general accuracy at all ranges attainable by rifles; its comparative independence of the excitement of battles; . . . its great endurance." Unlike the walking fossils who had staffed the ordnance department in Civil War days, the young bucks who worked there these days were intrigued by, not disdainful of, the advances in weaponry that were then sweeping the world.

The world: By the end of the 1860s, that, increasingly, was Gatling's most important market. The stakes had been raised. And for that reason, the fact that the U.S. military now recognized the value of his gun most likely didn't bring the kind of grand, crowning satisfaction it would have brought during the Civil War, back when Gatling was unknown, back when he believed with a young man's passionate certainty just what it was that he had created, the *momentousness* of it—but could not, alas, get the military men to realize it, too. The U.S. Army would end up purchasing a great many of his guns throughout the second half of the nineteenth century, but by that time, Gatling had moved on.

He had moved on to, among other places, England and France and Russia and Germany and Turkey. He began to travel regularly to those places, or send the sales representatives in his employ to those places, in order to show those heads of state—rulers who were nervous, as always, about what their neighbors might be buying in the way of weaponry—just what the Gatling gun could do. In 1869, Gatling negotiated the sale of the British rights to make his guns to Sir William George Armstrong, the private British arms merchant who had sold weapons to both the North and the South during the Civil War; arms dealers, as the world was learning, did not consider themselves bound by pesky ethical niceties. Gatling was a salesman as well as an inventor; he had come up with the product, and now he was going out on the road.

He was doing, that is, exactly what he'd done in the 1840s, when, fresh from patenting his seed planter, he had traveled from river town to river town on steamboats, talking to anyone who'd listen, and pushing, and *pushing*. Now it was a machine gun and world capitals and ocean liners, but it was still the same thing. He was older, heavier, the hair whiter, the step slower and more deliberate; yet he was still what he had always been, still untroubled by the fact that he had made a great fortune from blood and destruction. He was still the same man. Wasn't he?

His courtly, distinctive signature—"R.J. Gatling"—was comprised of elaborately linked swirls. To create it, the pen would not necessarily have had to lift from the paper. The entire name could have emerged from a single practiced stroke: The eastern tail of the *R* swells into the western edge of the *J*, and the *J* glides smoothly into the first *G* of "Gatling." The writer had a habit of adding a curlicue beneath his name, then bisecting it with two quick vertical marks, as if to say: *So there.*

The documents bearing this signature were, after 1870, officially filed away in the Colt Patent Firearms Company archives in Hartford, Connecticut. That was the year Gatling sold his patents for Gatling guns to Colt. Gatling retained the title of president of the Gatling Gun Company, but that company would be absorbed by Colt. Assigning the rights to manufacture what one had created to a large and established firm was a common and logical way for inventors to cash in. Gatling had done it in the 1840s with his seed planter, going from state to state, selling the manufacturing rights to other businessmen. When it came to a product as expensive and as labor-intensive to create as a Gatling gun, the Colt Company was a natural choice; it was a prosperous, like-minded firm that possessed the factory and staff to turn out the intricate weapon to Gatling's exacting specifications.

The same year, 1870, Gatling and his family followed his gun to Hartford. Richard and Jemima and their twelve-year-old daughter Ida moved from Indianapolis. It was here in Hartford that their two sons, Richard Henry Gatling and Robert Barnes Gatling, would be born, in 1870 and 1872. Henceforth Gatling guns would be made in the famous factory established by the late Samuel Colt along the Connecticut River, the factory that stretched out beneath the dark blue dome studded with gold-painted stars, a dome upon which a statue of a rearing colt added the final touch of muscular glamour. The first statue had been destroyed in an 1864 fire; the replacement zinc statue topped the structure from 1868 to 1988. Weapons of death were made daily here, but you could sometimes forget that fact when marveling at the magnificent colt, a broken staff in its mouth, the lifted hooves scraping the heavens. Beauty and danger: The Colt factory crackled with both.

Eighteen-seventy was also the year that Colonel James Wolfe Ripley, Gatling's old nemesis in the U.S. Army Ordnance Department during the Civil War, the man who never met a new weapon he couldn't reject, ignore, disparage, or sabotage, died. And he died in Hartford, the city to which he had decamped after retiring from military service in September 1863. Thus whether the two men knew it or not, their paths briefly intersected one last time. Ripley and Gatling, the bureaucrat and the businessman, a pair of trajectories heading in opposite directions: one falling, the life shortly to conclude, the life closing in on itself with the slow, muffled grace of impending oblivion; and the other one rising, vivid, his days now expanding with international travel and constant challenges, the hectic years shortly to blossom with opportunity. For an instant the lines crossed. The place where they crossed was Hartford.

Hartford. It wasn't just a city of commerce; it was a city of culture as well. America's Industrial Revolution was born here, but so was part of its aesthetic heritage. Hartford was the home of Mark Twain and Harriet Beecher Stowe and Frederick Law Olmsted. It was a city of vivid commercial might and great creative vitality, a city of smokestacks and poetry. Famous in their day—the late eighteenth century—was a group of poets known as the "Hartford Wits." Noah Webster published his first dictionary in Hartford. Charles Dickens, after a three-day visit in 1842, called it a "lovely place," adding, "The town is beautifully situated in a basin of green hills." Twain waxed positively rapturous on its behalf: "Of all the beautiful towns it has been my fortune to see," he wrote in an 1868 magazine article, "this is the chief. . . . Each house sits in the midst of about an acre of green grass, or flower beds or ornamental shrubbery . . . and by files of huge forest trees that cast a shadow like a thunder-cloud. Some of these stately dwellings are almost buried from sight in parks and forests of those noble trees. Everywhere the eye turns it is blessed with a vision of refreshing green. You do not know what beauty is if you have not been here." The scribbling cynic was completely besotted by Hartford. And why not? Along with the stately homes of the wealthy—Hartford citizens possessed more money per capita than those of any city in the United States—there was an elaborate system of public parks. Hartford was the first American city whose public spaces were lit completely by electricity. To Twain, comparison with other cities of his acquaintance was preposterous; next to Hartford, he wrote, St. Louis was "a muddy, smoky, mean city to run about in."

Between 1850 and 1860, Hartford's population doubled, boosted by a large number of German and Irish immigrants who sought employment in the great manufacturing industries that lined the Connecticut River Valley. The flagship of those concerns was Colt's Patent Firearms Company. Even though Samuel Colt had died in 1862, his widow, Elizabeth, had made a series of shrewd business decisions, and the Colt works were thriving as never before. Other factories, other businesses, also prospered here. By 1876, *Scribner's Magazine* called Hartford "the richest [small] city in the United States." The city was "a Silicon Valley of its day."

The country as a whole was in a dandy mood. Business was good. Business was very good. And business was most all that mattered to Americans, as Henry Adams, that ruefully self-aware chronicler of his countrymen, noted: "The world, after 1865, became a banker's world."

Just three years after Gatling and his family moved to Hartford and settled in at 27 Charter Oak Place, however, the trap door beneath the American economy fell open, the way it seemed to make a habit of doing with distressing regularity. There had been a panic in 1837, then another in 1857, and now there was a desperate crunch in 1873. Other panics would occur, right on schedule, in 1884 and 1893. Among the reasons for these familiar but always frightening downturns was the lack of a central bank in the United States. The central bank had been shut down in 1836 and would not be reinstated until 1913. This enabled wildly unchecked swings in the economy. There was "no official lender of last resort, no federal recourse in times of acute turbulence or panic," writes Jean Strouse. "America's antiquated banking system had been devised before the Civil War for a de-centralized agricultural society." And here the nation was, moving rapidly toward a complicated, highly centralized, industrial society.

Few spots embodied the contradictions, the vexing riddles of the expanding-and-contracting American economy, its dreams and its nightmares, better than Hartford, where Gatling and his family lived for twenty-seven years. Hartford was a lovely place, a place of abundant trees and flowers, of serenely gracious homes, but it was also the center of the nation's modern armaments industry. America saw itself as a symbol of freedom, of belief in self-determination, yet the weapons its factories turned out were used around the world to strengthen the grip of colonial empires. Guns epitomized the American system of manufacture, inspiring the development of the precision instruments that would revolutionize the business world: the drop hammers, the jigs, the boring machines, the machine tools such as lathes, drill presses, and grinders. The instruments that made the modern world possible also ended up creating the weapons that would be used to tear it apart. Hartford, a garden spot, an enchanted-seeming realm of beauty and refinement, also was the place from which emerged a steady supply of Colt pistols and Gatling guns. Hartford had "the broadest, straightest streets," claimed Twain, whose family relocated there a year after the Gatlings did, "and the dwelling houses are the amplest in size, and the shapeliest, and have the most capacious ornamental grounds about them." Many of the biggest and most beautiful of those homes were bought with money from guns.

During the heyday of the Colt Armory, the steam whistle blew each morning at 7 A.M. and could be heard at least twenty miles away, summoning employees to their eleven-hour workdays. The Colt factory was the largest workshop in the world. And it was run safely as well as profitably. A Gatling gun, in fact, caused the only recorded death in the company's long history in the nineteenth century; a workman named John Kallaher was killed in 1870 while testing the gun in a shooting gallery.

Arriving at the Colt Works in the morning, Gatling would have seen "a great range of tall brick buildings, and on every floor . . . a dense wilderness of strange iron machines that stretches away into remote distances and confusing perspectives—a tangled forest of rods, bars, pulleys, wheels, and all the imaginable and unimaginable forms of mechanism," according to Twain's account of the place. "There are machines to cut all the various parts of a pistol, roughly, from the original steel; machines to trim them down and polish them; machines to brand and number them; machines to bore the barrels out; machines to rifle them; machines that shave them down neatly to a proper size, as deftly as one would shave a candle in a lathe; machines that do everything but shape the wooden stocks and trace the ornamental work upon the barrels."

What made American guns the best in the world wasn't just their design, as innovative as that might be. It was the precision and cost-effectiveness with which they were made. Interchangeable parts could only be "interchangeable" if those parts were uniform. And the American system of manufacture, with the machine tools pioneered at the Colt Armory, found its glorious zenith in the making of firearms. Had there been no American arms business, there would have been a very different American Industrial Revolution—if there had been one at all.

• • •

From nations around the world, orders for Gatling guns rushed in from kings and prime ministers and generals. Back in the United States, though, it was not only military men but also police officers and wealthy private citizens who desired them. By the 1870s, mine owners and railroad tycoons had discovered that Gatling guns came in handy for keeping discontented workers in line. Gatling guns were a coolly efficient way of demonstrating that the men in charge meant business. Gatling guns were routinely hauled out to break up labor riots in an especially violent span of months in 1877. In July of that year, Gatling guns were brandished by the New York Militia against workers striking the Erie Railroad. In 1880, the secretary of the Treasury ordered the purchase of three Gatling guns for installation atop Treasury buildings as a show of force. Before the end of the nineteenth century, militias in Kansas, San Francisco, Connecticut, Tennessee, Pennsylvania, and Ohio bought Gatling guns to show off to striking workers or unruly citizens in general. Even if it wasn't fired, a Gatling gun made a powerful statement, and commanders of Western forts liked to park them, polished and solemn, just outside the main gates. A Gatling gun was like a "Beware of Dog" notice. It made potential troublemakers hesitate, then back away. That, of course, had been Gatling's idea right from the start; that was his ethical rationale for creating it. That was how he justified it to those who wondered how a gentle, good-hearted man could unleash such a horrific weapon upon the world: It could actually prevent bloodshed, he claimed. It could limit, not instigate, excessive casualties. Some people quite understandably tended to doubt Gatling's sincerity when he posited that, but in many cases, actual use of the guns was proving his point.

The fact was, Gatling guns just looked mean. Mean and menacing and unforgivably fierce, yet sleek, too, and as precise as surgical instruments. In the world of the late nineteenth century, where labor-saving devices had captivated the public imagination, Gatling guns were the ultimate labor savers: They made military action more cost-effective than ever before. They were solid emblems of American might, of American technological prowess, that also sported a nasty streak. Even as other machine guns entered the market—by the 1880s and 1890s, Benjamin Hotchkiss, Hiram Maxim, and John Moses Browning, among others, had introduced their versions—the Gatling gun was what persisted in the public imagination, what loomed large and dangerous-looking. Gatling guns weren't the only machine guns around, but when people heard the phrase "machine gun," they envisioned the Gatling. It was the Gatling gun that people thought about when they thought about the new machinery of death.

Gatling guns were symbols, but they were also more than symbols. A half dozen times between 1874 and 1878, Gatling guns were fired at Native Americans who resisted the takeover of their homelands. Major General Oliver Otis Howard employed two Gatling guns as his troops chased Chief Joseph and the Nez Percé north toward Canada. "Employed under conditions ranging from the parched plains of the Anadrako Agency to the deck of a steamer on the Columbia River, Gatlings generally performed satisfactorily," wrote military historian David Armstrong. "During the Red River War, Lt. Col. Thomas H. Neill, Sixth Cavalry, used Gatling fire to prevent hostile Cheyenne Indians from entrenching themselves within rifle range of his position near Darlington Agency."

In 1876, several Gatling guns were offered to a simpering, arrogant cavalry officer named George Armstrong Custer just before his encounter with a band of Sioux in a place called the Little Big Horn. His first impulse was to accept them. Later, he turned them down. His refusal to take along the Gatling guns—probably a sound decision, many historians say, because the gun carriages of that era would not have fared well across the craggy terrain—ended up reinforcing the legend that grew up around Custer's defeat. It seemed to confirm Custer's haughtiness and overconfidence, the preening hubris that led to his tragic miscalculation. The Gatling gun was proving to be as significant for where it wasn't used as for where it was.

Despite a reputation as a brutal persuader, the Gatling gun still was hounded by prejudice and misinformation. Tests conducted by U.S. military officials proved again and again that the guns were reliable and accurate, but rumors persisted that they were "clumsy" and "frightening," as Evan S. Connell calls them in his otherwise estimable account of Custer's fall, *Son of the Morning Star: Custer and the Little Bighorn.* It is true that one of Custer's officers records that a horse-drawn carriage with a Gatling gun had overturned in the weeks before the battle, injuring a trio of soldiers. That, however, is hardly the fault of the guns themselves. "These weapons were invented in 1861 and had been very little improved since then," Connell adds. "They frequently malfunctioned." Neither assertion is true. Gatling worked on the gun continuously after obtaining the original patent in 1862, fixing problems that users encountered. In fact, he never stopped improving it.

Between 1862 and 1876, the year of Custer's defeat, Gatling redesigned the gun at least six times, adding an arched frame front, a drum magazine, and an automatic oscillator; shortening the bolt and breech housing; raising the walls on the feed hopper for better support of the magazine; moving the front and rear sights; attaching a crank lock. Gatling guns rarely "malfunctioned," as Connell charges, according to the reports of reputable ordnance officers. In fact, after tests of Gatling guns in the summer of 1863, an officer wrote of the weapon: "Mechanical construction is very simple, the workmanship is well-executed, and we are of the opinion that it is not liable to get out of working order." Earlier that spring, Navy Lieutenant J. S. Skerrett had informed Rear Admiral John A. Dahlgren that the Gatling gun "has stood the limited test given it admirably; has proven itself to be a very effective arm at short range; is well constructed and calculated to stand the usage to which it would necessarily be subjected." Another report after a U.S. Army trial was blunter: "All parts of the gun work well." But the backstairs gossip about the Gatling gun continued, and even though these accusations were groundless, they still inched their way into histories of the period, and they have dogged the reputation of Gatling guns ever since. The problem arose early: Armies are run, and always have been run, by romantics, by men who must believe in the noble mission. How else to justify the pain and waste and futility of war? There was nothing romantic about bolts and breeches and turning ratios. It was an impediment that Gatling and other inventors would come up against repeatedly. "Certainly they acknowledged that soldiers got killed by firearms," notes one historian, writing about the typical run of officers, "but they were never prepared to admit that advances in technology had reached such a level that the staunchest assault by the best of troops could be brought to nothing by modern weapons."

In a letter to a business associate dated October 14, 1868, with a Berlin postmark, Richard Gatling wrote, "I returned to this city a few days ago, having been since I last wrote you, at Stockholm & St. Petersburg. . . . Trials of my guns will also soon take place at Munich. . . . I still think that most of the governments of Europe will, ere long, give orders for my guns, for surely they have no rivals." Soldiers in the field may have been reluctant to embrace the big new guns with their deadly mechanical effectiveness, but heads of state showed no signs of such touching nostalgia. Guns were guns, and in an uncertain world of ambitious enemies and shifting alliances, guns were indispensable. As the nineteenth century continued, weapons began to acquire even greater acceptance as just products rolling off assembly lines: products, not symbols requiring emotional responses or political justifications. Weapons had long been bought and sold around the world; as early as 1849, Samuel Colt was peddling his guns in places such as Great Britain and

Turkey. Yet increasingly they were regarded as simply more wares for the auction block, indistinguishable from objects such as ships and apple peelers and sewing machines.

At the same time, though, the companies that manufactured armaments did attain a unique status. Their inventory was composed of products after which nervously paranoid nations hungered and hustled, so the arms makers enjoyed a special position. In the 1870s and 1880s, "international arms manufacturers were tacitly recognized as international powers," one historian noted, hence "dealt directly with sovereigns." In an admiring 1879 profile of Gatling, a writer enthused, "Great states constitute his clientage, and he deals direct with kings."

And even kings were forced to kneel to arms firms. When Krupp sold guns to both sides of the war between Russia and Turkey in 1877–78, the company's public relations staff garnered testimonials from each combatant, as William Manchester reported in his history of the German weapons behemoth. Arms firms were virtually extrasovereign. They did not have to follow the same rules that everyone else did. Because when wars threatened, the leaders—and the people they led—easily forgot about the niceties of corporate practices, about ethics and fair play. They wanted the weapons. And they wanted them fast, and they wanted a lot of them.

It was not, however, an easy or simple business, as Gatling and his competitors had discovered. Demand for products might be intense, but it was sporadic, "being heavily dependent on the erratic movements on the graph of international tension," wrote business historian Neil McKendrik in his introduction to a study of Vickers Brothers, a British arms firm that prospered in the late nineteenth and early twentieth centuries. The customer base was quite small—composed as it was of nations, not individuals—and the amount of capital investment dauntingly large. It was that rare case of a monopoly that existed among the customers, not the suppliers.

Armaments manufacture was a matter of competitive research and development, of innovation and profits and sales. What had been the work of private artisans—making and selling weapons of death—now was a series of lines on the balance sheets of famous and admired companies. You could almost forget, for long stretches, that these immense and heavily advertised and actively promoted weapons actually were intended to maim and kill; at times they could seem like simply more gleaming products emerging from the miraculous machinery introduced by the Industrial Revolution, like cook stoves or engine parts. There was little shame attached to armaments sales. No long nights of the soul for arms merchants, wondering if their weapons were hastening the world's self-destructive course. "They lived in a period when, by and large, the armament business was accepted as a heavy industry, much like any other, if technologically a good deal more specialized," wrote a biographer of Vickers Brothers. "When they dealt abroad, they accepted the standards of the marketplace in which they traded, as did all other commercial concerns. They were not 'merchants of death' . . . and could not consider themselves this way; but they were good businessmen."

Five multinational firms were locked in vigorous competition with each other during this period to supply the world's heavy-duty weapons: Schneider, Armstrong, Vickers, Mitsui, and Krupp. Arms were no longer the work of small scale artisans working in isolation. Arms were products. They required skilled businessmen, complete with elaborate manufacturing plants, to keep them coming. By the end of the nineteenth century, "Warriors didn't fight in the saddle with pikes, halberds, and epees any more," Manchester noted; so "foundries, rolling mills and smokestacks would be far more appropriate symbols" of the weapons business. And Fritz Krupp, Manchester added, was at once "an industrialist, gunsmith and diplomat."

Nations anxious about their place in the world's hierarchy of power realized that armaments were key. Russia was a good example. After its crushing defeat by Great Britain and France in the Crimean War in the mid-1850s, Russia embarked on a relentless campaign to update its military. That meant compiling a giant shopping list for the world's most innovative arms. It would buy its way into safety. "Official [Russian] policy devoted great effort to equipping the Tsar's army and navy with the latest and most efficient weapons, even when they had to be purchased abroad from Krupp or Armstrong," according to historian William McNeill. "Russia, in fact, ranked among the very best customers for both these firms from the 1860s onwards."

More than any other business, the arms trade, as it developed between 1840 and 1880, created the modern global marketplace. "Cheap machine-made goods and cheap machine-based superiority of armed force were both available for export, and exported they were," McNeill writes. "As a result, the world was united into a single interacting whole as never before. World markets reached across all existing political boundaries. . . . More than a century later we remain the heirs of this achievement." The arms business was a *business*, and arms makers neither apologized nor explained. Profits, after all, were most definitely profits.

Just how respectable was the world of armament manufacture? In 1898, Fritz Krupp gave a lavish party at a Berlin hotel. The sumptuous place settings included "a miniature ship decorated with . . . a tiny gun loaded—not with lethal shot and shell—but with violets or other flowers." Likewise, at the wedding reception for Samuel Colt and his bride Elizabeth in 1856, the cake was "six feet high [and] trimmed with pistols and rifles made of sugar." Weapons could be downright whimsical.

Creating his guns had challenged Richard Gatling's ingenuity. Selling them would test him in altogether different ways. For weapons sales did not allow for rest or complacency. Arms makers such as Gatling faced constant threats in a cutthroat international marketplace. Rival inventors stayed just as busy as Gatling did; machine guns with names such as Hotchkiss and Maxim and Gardner and Lewis and Nordenfelt were seriously cutting into Gatling's business.

Stung by reports that other brands of guns could outperform his, Gatling bought a pugnacious ad in the August 27, 1881, issue of the *Army and Navy Journal.* He had it reprinted in a British magazine the same year, the only change being a switch of the amount of the bet to pounds sterling:

> *A Notice*
>
> THE GATLING GUN
>
> Many articles have recently appeared in the press, claiming the superior advantages of the Gardner and other machine guns over the Gatling gun.
>
> In order to decide which is the best gun, the undersigned offers to fire his gun (the Gatling) against any other gun, on the following wagers, viz.:
>
> First, $500 that the Gatling can fire more shots in a given time, say one minute.
>
> Second, $500 that the Gatling can give more hits on a target, firing, say, one minute—at a range of 800 or 1000 yards.
>
> The winner is to contribute the money won to some charitable object.
>
> The time and place for the trials is to be mutually agreed upon.
>
> R. J. Gatling
> of Hartford, Conn.

There are no reports that anyone ever took up his challenge. But if they had, Gatling surely would have followed through; he had just that kind of unassailable faith in his invention, that conviction, that deep belief.

Competition from other guns, however, wasn't his only headache. The cost of doing business also rose steadily; among other reasons, ammunition was always being improved, and updated ammunition necessitated updated arms. Smokeless powder was introduced in 1886; a short time later, copper-jacketed, lead-core bullets came along. The first Gatling guns were designed for paper cartridges. When there was a breakthrough in bullet design, or when the American military decided to change the bore diameter of its weapons—and thus the caliber size of the bullets—Gatling had to alter his guns. The inventor later estimated that the overall cost of retooling Gatling guns to accommodate the military's changes in ammunition size easily topped $1 million.

Thus the fortunes of Richard Gatling and his gun oscillated, as fortunes do, and as they especially did in the last half of the nineteenth century, when business cycles went up and down like mood swings. Until things settled down, there would be a wildness to the business world, a patternless unpredictability. And the arms business, which stretched across oceans and borders, was especially subject to volatility. Gatling was rich and poor and rich again; then poor. "I regret to say that my gun business has not been so good for the past year as usual," Gatling wrote to his niece Rebecca in February of 1895. Conniving inventors in other countries had begun to swipe the design innovations of the Gatling gun to avoid paying royalties; copycat piracy was common then, just as it is now. Such was the double-edged nature of fame: The Gatling gun had become so popular that it was worth stealing. And yet that very popularity would ensure the Gatling gun's future success. It was so illustrious that many people couldn't wait to get their hands on one. A twenty-year-old Edgar Rice Burroughs, who later would create Tarzan and then send him swinging through a series of enduringly popular novels, became a Gatling gun instructor at the Michigan Military Academy on July 4, 1895. Gatling guns easily won the hearts of the adventurous.

While his business suffered sharp downturns over the years, while it was always a challenge, Gatling still made deals with armies around the world. Among the countries that purchased Gatling guns were Argentina, China, Denmark, Egypt, France, Great Britain, Holland, Italy, Japan, Korea, Mexico, Morocco, Romania, Spain, Sweden, Switzerland, Tunis, and Turkey. Yet once again, success carried a sinister implication: Gatling guns became the weapon of choice for British forces determined to enforce colonial rule in Africa.

The Gatling gun, that is, had become something its inventor could probably not have foreseen: a tool of racism. Newspaper and magazine accounts of its use in the last half of the nineteenth century make that distastefully clear. When the British fired Gatling guns at the Ashanti people in Africa, an editorial warned, "The savage does not fight by the rules of modern tactics." Apparently most Ashanti surrendered rather than contend with the peculiar-looking weapons, leading a correspondent for the *Army and Navy Journal* to gloat, "We are not surprised that the Ashantis were awe-struck before the power of the Gatling Gun. It is easy to understand that it is a weapon which is specially adapted to terrify a barbarous or semi-civilized foe." Gatling guns also were used by the British in the Zulu Wars in 1879. A correspondent wrote, "When all was over and we counted the dead, there lay, within a radius of five hundred yards, 473 Zulus. They lay in groups, in some places, of fourteen to thirty dead, mowed down by the fire of the Gatlings, which tells upon them more than the fire of the rifles." Gatling guns were used by the British in Egypt and the Sudan. They were unleashed in northern Nigeria. They were employed anywhere, in fact, that a show of deadly force was required, and where the lives of the enemies to be subdued were regarded as not quite worthy of honorable traditional combat, of a fair fight and even odds. "Round whisked the Gatlings, r-r-r-r-rum, r-r-r-r-rum!" went a breathless account in the *Army and Navy Gazette* of an 1882 battle in Africa. "That hellish note the soldier so much detests in action, not for what it has done, so much, as for what it could do, rattled out. The report of the machine guns as they rattle away rings out clearly in the morning air."

How strange it was that as the nineteenth century unwound the Gatling gun became, in effect, an enforcer of racism. Richard Gatling had invented it, after all, to help the North win the Civil War, a war fought principally to end slavery and, by extension, the scourge of racism. Yet the Gatling gun's most conspicuous use in the late nineteenth century was to enforce the racist-inspired subjugation of African populations. "With machine guns in their armoury," wrote historian John Ellis, "mere handfuls of white men, plunderers and visionaries, civilians and soldiers, were able to scoff at the objections of the Africans themselves and impose their rule upon a whole continent." People who looked different from white Europeans were regarded as subhuman, as an undifferentiated mass that could be cut down by machine-gun fire like so many stalks in a field. The idea of an individual life—or death—being of consequence had no standing here, here where the skins were dark and the customs odd. "Machine guns were lengthily regarded, notably by the British, as improper instruments of war, save for quelling unruly mobs, the yelling and fanatical forms of enemy who had to be suppressed," an arms historian wrote. "That was not proper warfare. It was bringing into line some awkward enemies who needed to be taught some rules." The attitude is reminiscent of the way the American military employed the Gatling gun during this period: It was fine to use against "savages" in the West, or against striking laborers, many of whom were immigrants. The racist assumptions behind machine-gun use would persist even through the opening salvos of World War I; a German soldier wrote that many of his comrades "regarded the machine gun as a weapon for use against Hereros and Hottentots." But not as a weapon to use against one's own kind.

Was Gatling a racist? His father had been a slave owner, and Richard Gatling and his wife were given a slave named Rachel Stepney as a wedding present by the Sanders family. They later freed her, according to Gatling's grandson, who wrote that his grandparents "were not in favor of slavery and its system." But it was possible to be opposed to slavery and still to harbor racist ideas. No one can know what Gatling, in the privacy of his thoughts, really believed about race. Yet the Gatling gun indisputably functioned as a kind of iron fist of racism in bloody, one-sided battles.

He could sometimes appear preoccupied, distracted, withdrawn. When Richard Gatling was nervous and anxious, he had a habit of pulling at his beard. His brothers had the same habit. It was a trait of the Gatling men: They were thinkers, contemplative sorts. But those who knew Richard Gatling best also said that he would happily pry his head out of the clouds when he needed to, and most often it was to pay heed to his family, which he loved dearly and without reservation.

The Gatlings had moved easily, seamlessly, into Hartford society, becoming habitual churchgoers and smiling guests at charitable and cultural events. Their swift acceptance by longtime residents was naturally a tribute to how they were—polite and interesting—but it also said something about *who* they were: Richard Gatling was rich and famous. In a socially ambitious city such as Hartford, that surely would tend to enhance one's desirability as a friend and neighbor. It was no surprise, then, that twenty-two-year-old Ida Gatling's wedding on October 14, 1880, was a starred date on the Hartford social calendar.

The wedding was held at the South Baptist Church in Hartford, where the groom, Hugh Owen Pentecost, served as pastor. The lavish wedding was, according to a newspaper account, "one of the most fashionable events of the season." A prewedding photograph of Ida Gatling, an image rinsed by time and historical distance into a soft gray portrait with no sharp edges or clear details, shows a shy and somewhat apprehensive-looking bride glancing demurely over her left shoulder. In a separate photograph, the groom, who later would leave the ministry and become a corporate lawyer in New York, sports protruding ears, a fleshy lower lip, a strong chin, and round spectacles that give him a look of pinched and dour earnestness. Dark, wavy hair is pushed well back from the lined forehead, like a stray and troubling lascivious thought.

For the reception, the Gatling home was transformed into a veritable hothouse: "Banks of flowers have been placed on all the mantels and mirror cases," runs the breathless newspaper description, "and in front of a large mirror in the reception room a massive horse shoe of beautiful flowers has been placed. The chandeliers and other available points have been taken advantage of, and large baskets have been placed in each room. Flowers abound all through the mansion."

For all of his great fortune, however, and for all of the material comforts and social satisfactions that large sums of money could bring, including a fancy, flower-draped wedding for his beloved daughter, Richard Gatling was not just another successful businessman amid America's rising prosperity. He was something more. He was different because of what he made, and because of where he came from. He was the living embodiment of the journey the country had undertaken, shifting from a rural agrarian economy to an urban industrial one in only a few decades. He had moved from South to North, just as the money had. Before the Civil War, the South possessed almost a third of America's wealth; five years after Appomattox, that percentage had dropped to 12. The South's per-capita income in 1860 was about two-thirds of what Northerners earned. At war's end, that number was reduced to two-fifths.

So he had bet correctly, Gatling had, with both his life and his inventions. He had rightly intuited the momentum of the nineteenth century: toward bigger things, toward mass scale. He was a middle-aged man now, well settled into his fifties. He looked every minute of his years, with his spreading midsection and thinning hair, hair long gone to white, and the rhyming white beard. Yet at times you still could glimpse within him the lean shadow of the vigorous young man, of the kid who used to watch the boats on the Meherrin River, dreaming of making them better and faster.

The river was much in his thoughts these days. He had returned to the family farm several times over the decades. A year before Ida's marriage, he had gone back for the saddest of reasons. His brother, James Henry Gatling, a bachelor who had stayed on the family farm, had been murdered by an enraged neighbor over a petty dispute. Richard Gatling attended the killer's trial in Winton, the county seat of Hertford County, and heard the guilty verdict. And then there was the estate to be dealt with, all the lawyers, the property, the fuss. Gatling's letters in these years to his North Carolina relatives are filled with the business of deed transfers and crop estimates and weather. His elegantly curved signature often follows a long analysis of the chopping-up of the family property; a piece here, a piece there. He never shirked his duty. Richard Gatling always came home when he was needed. He would sometimes take his sons, but not his wife; there is no record of Jemima Gatling ever visiting her husband's boyhood home.

In August of 1884, he made another family trip, on yet another unfortunate errand. From the Queens Hotel in Toronto, Canada, Gatling wrote a letter to his niece, Rebecca G. Peebles, informing her that his younger brother William had died near that city after a long illness. Gatling had crossed the northern border to see to the arrangements. "It is needless to say his death has filled my heart with great sorrow and grief," he wrote, "but, we must all sooner or later pass from earth, and I hope and trust in the spirit world, we shall be in a better world than this, where there will be no pain or sorrow, and be with our kind Maker in heaven." Gatling added that he had decided to have his brother buried in the Masonic cemetery in Toronto, "rather than take the body to N.C. for burial which would be almost impossible to do in this warm weather—moreover, my dear wife is in feeble health at this time, and I could not be from home for so long."

William Gatling had not enjoyed the same worldly success as had Richard. He had sought his fortune in Canadian silver mines, but at his death he left, Richard Gatling had discovered and now informed his niece, "a case of books & his gold watch and a few other articles of not much value," along with a few shares of stock in those mines and a few acres of land in White Cloud, Kansas. The tone of Richard Gatling's letter was uncertain, even a bit defensive, as if he anticipated family criticism: "I have done, My Dear Niece, the best I could under the circumstances & hope all the relations will be satisfied with what I have done." Richard Gatling and his brothers had followed different rivers as they moved beyond their youth, into the world at large; but everything eventually would come back to one river, the river of the past, and it ran, as it always does, from darkness into light and then into darkness again.

In 1897, after nearly three decades in Hartford, with their children grown and gone, with the Gatling gun business shakier than ever because of competition from newer weapons, Richard and Jemima Gatling moved to New York City. They would live there with Ida and her family, which now included a young daughter. A few months after Richard and Jemima left Hartford, their pastor back at the South Baptist Church, Frank Dixon, wrote to Jemima to let her know just how fondly the Gatlings were remembered: "If you knew how sadly we miss you from the church and from our lives, your heart would incline you to come back to Hartford, I know. . . . No one has taken your place and no one can. . . . We have quite a number of acquaintances in our church here, but very few indeed whom we can regard as friends."

It might have been polite boilerplate, but it has a ring of sincerity. Neighbors had always found Richard Gatling amiable and kind. Not at all like the fiery, truculent Colt, the city's other famous gunmaker, whose legend had only seemed to swell with his early and mysterious death in 1862. Colt was much more like what you'd expect a gunmaker to be: ruthless, blunt, profane. Yet given the business in which both men made their fortunes, Colt's personality is much easier to understand and accept. Gatling is the oddity. It is difficult to reconcile the man who created the Gatling gun with the loyal husband and gentle father, to reconcile the canny, competitive arms merchant with the decent, peace-loving citizen. And the same paradox had begun, by the closing decades of the nineteenth century, to crimp the edges of America itself, as the nation became a financial and military empire. American citizens believed, after all, "that heaven intended it to free the world, not rule it," writes historian Niall Ferguson. Its discomfort with its new status as global powerhouse was beginning to show; it was shortly to become, in Ferguson's phrase, "an empire in denial." Was America the world's friend or its master? Was the Gatling gun ultimately a force for peace and justice, as its inventor claimed, or just another weapon of death that brought profits to its makers? On chilly Hartford mornings, when Gatling stood on the broad wooden front porch of his home at 27 Charter Oak Place just before he left for work, he would have heard, behind him, the sounds of a busy household—his wife and his children, beginning their day—and he would have seen in front of him, rising in the mist from the Connecticut River like the approaching messenger from some distant army, the rearing horse on the great Colt dome. Poised between hearth and lathe, between good intention and lethal outcome, was Richard Gatling, husband and father and Christian. And gunmaker.

He never gave up. He never stopped inventing, which is another way of saying he never stopped dreaming and hoping. He was stubborn. But it was a good kind of stubbornness, the kind that keeps you going when others doubt you. "You could not impose on him, for he was very firm and strong about his 'rights,'" wrote his grandson, John Waters Gatling, who was born after Richard Gatling's death but who had absorbed the family consensus about the inventor. "Differences were to be settled by negotiation, and people were to be free, unregimented, and individuals." Late in Richard Gatling's life, with many of his dreams in shreds around his feet—most of his fortune had been lost to ill-timed real estate speculation—Gatling kept inventing. He adapted the Gatling gun for a belt feed instead of a drum magazine. And he moved beyond guns. He patented a bicycle on May 8, 1894. He patented a device to control wagon reins on October 10, 1899. He patented a new kind of flush toilet on February 26, 1901. His last patent was granted on July 22, 1902, and that was the one upon which he had pinned his hopes; that was the big one.

His steam-driven motor plow was going to change the world. It would make people forget all about the gun. It would, he told a St. Louis newspaper, "accomplish on the farm what his gun did on the field of battle." Gatling understood farming and farmers; he had made his first fortune in agriculture. Better days, he knew, were ahead. He shared his country's golden optimism. It was the deepest part of him. It had not aged as the rest of him had aged.

• • •

And so in 1902 Richard Gatling and his wife Jemima relocated temporarily to St. Louis, just until he could get the new company up and running. The eighty-four-year-old inventor was starting from scratch: a new business, a new kind of device. A steam tractor. He'd be fighting skepticism, if not outright indifference, and some firmly entrenched financial interests—but he'd done it all before, hadn't he? He had done it once. He could do it again. He lived in a land of second chances. And besides, he had decided that the best place to manufacture his plow was St. Louis, site of his first great success: the seed planter. Yes, yes, that was almost sixty years ago, but no matter. Richard Gatling knew machines. He knew markets. He knew what he was doing.

Problems accumulated. Setbacks and frustrations mounted. Gatling couldn't find a decent manufacturer. He needed capital, and that meant he needed to talk up his invention, but he was afraid that if he talked *too* much, he'd give away crucial details and someone would steal his idea. His heart, moreover, was weak; physicians had warned him about exerting himself. A bout with the flu left him gaunt and frail.

Back in New York, Ida read, with increasing dismay, her mother's letters describing the situation in St. Louis. Ida dashed off a telegram to Jemima in early 1903, hoping to provide a bit of cheer: "You are lovely as the month of May. Ida."

Finally, a desperate Gatling asked his oldest son for financial help. Richard Henry Gatling, by now a corporate lawyer in New York, obliged, and Gatling acknowledged the gift in a brief note:

My Dear Son Richard,

Your letter of the 17th enclosing check for $500 came only to hand and in reply, I am at a loss for words to express my gratitude for your kindness. The amount will I trust enable me to get my plow completed & tested. I would have had my plow finished had it not been for the slowness & lying of parties here to do the work as they had by contract agreed to do. . . . I speak of these things so you might know what I have had to contend against. . . . I shall certainly push smartly with all possible speed, for I am as well as Dear Mother extremely anxious to get home and see you all. . . .

Ever your grateful & affectionate father,
R. J. Gatling

P.S. So again thank you for your great kindness in assisting me in my time of need.

The letter is riven with obvious emotional pain, but it is also suffused with a kind of golden autumnal light, with the rightness of a natural reciprocity: First parents help and guide their children, and then, sometimes, the child is called upon to help the parent. The steam-motor plow was Gatling's final dream. It was his last great adventure.

The Gatling gun, too, had one more glorious adventure left before it was superseded by the weapons its creation had enabled. It began on a sweltering hill in Cuba and eventually helped propel a blustery, bellicose young man of indomitable vigor and overweening ambition into the White House.

"As he turned on his heel a bullet struck him in the mouth and came out the back of his head, so that even before he fell his wild and gallant soul had gone out into darkness." Thus Theodore Roosevelt recorded the appalling sight of one of his men being cut down as they charged toward Santiago. "He stood up to salute," Roosevelt recalled of another comrade, "and then pitched forward across my knees, a bullet having gone through his throat, cutting the carotid."

On February 15, 1898, the U.S.S. *Maine* had exploded in Havana harbor. Certain that Spain had done the deed—although recent historical scholarship now tends to cast some doubt on that conclusion—Americans were outraged, ready to fight alongside the Cubans for their freedom from the Spaniards. No one was more eager to go than Roosevelt, forty, the son of wealth and privilege, who resigned his post as assistant secretary of the navy to gather a motley crew of soldiers—the "Rough Riders"—to fight in Cuba. These handpicked men "were trained by life-long habit to look on life and death with iron philosophy," Roosevelt recorded in *The Rough Riders,* his classic account of the struggle, published a year after the battle.

The Spanish-American War would be short, but not without peril. Through devastating heat, against an opponent equipped with smokeless powder and more modern artillery, the Americans rushed up San Juan Hill into a curtain of vicious firepower. Closing in on the summit, Roosevelt and his men paused. They heard a fierce sound like a great relentless drum. It caught the rhythm of their own madly accelerating heartbeats. "The Spanish machine-guns!" cried some of the Rough Riders in alarm.

But Roosevelt wasn't sure. He jumped up, realized that the sound emanated from land to the left, slapped his thigh, and yelled, "It's the Gatlings, men, our Gatlings!" a call that would resound in American military history. Lieutenant John Parker had arrived with his quartet of Gatling guns, providing covering fire for Roosevelt and the Rough Riders. The men gave a roar of enthusiasm and charged forth, energized by the hammering of the guns. "We saw much of Parker after that," Roosevelt wrote, "and there was never a more welcome sound than his Gatlings. It was the only sound which I ever heard any men cheer in battle."

Later, Roosevelt heaped more praise on Gatling guns and on Parker, who wielded them with strength and skill. "Indeed, the dash and efficiency with which Gatlings were handled by Parker was one of the most striking features of the campaign. . . . Parker's Gatlings were our inseparable companions throughout the siege." In the mopping-up operation that followed the decisive battle, "we usually got Parker to send a Gatling along, and . . . the Gatlings went, over any ground and in any weather." The guns, like the Rough Riders themselves, weren't delicate or soft or unreliable; they did the job. The Gatling gun had finally made its bones in what Roosevelt called "the blood-bought victory under the tropic skies of Santiago." It had become the thing that all great weapons must be: part metal, part legend.

Many more years would pass before the true implications of Gatling's technical achievement, of what he had created back in 1861, would be understood. The staggering significance of machine guns would be realized only when the descendants of the Gatling gun began to be classified as infantry, not artillery, weapons. Yet because Gatling guns originally were mounted on carriages, as were cannons and howitzers, few people saw the potential.

Roosevelt, though, had seen enough. His book was a veritable valentine to Gatling guns. And the book was a roaring success. It went through fifteen printings in his lifetime, it was quoted and memorized, and it helped kick-start the image of the twenty-fifth president as a bold, sturdy hero, more at home on horseback in the wild than in a parlor arguing with other politicians. He was utterly smitten by machine guns. He would be the first president to ride in a car and fly in an airplane, yet he found his true technological soul mate in the Gatling gun. The modern world was born.

CHAPTER SEVEN

"The World's Great Storm"

Bombardment, barrage, curtain-fire, mines, guns, tanks, machine-guns, hand-grenades—words, words, but they hold the horror of the world.

—Erich Maria Remarque, *All Quiet on the Western Front*

He grew used to the sight and smell of torn human flesh. He watched the men harden to the mechanical slaughter. There seemed to him a great breach of nature which no one had the power to stop.

—Sebastian Faulks, *Birdsong*

On what would turn out to be the last morning of his life, Richard Gatling walked the approximately four miles from his daughter's home at 249 W. 107th Street, an apartment building known as the Aberdeen, to the offices of *Scientific American*. He and his wife had returned to New York just a few weeks earlier; their son, Richard Henry Gatling, had gone to St. Louis to fetch them.

It was February 26, 1903. Gatling was eighty-five years old. He was the inventor of the most famous weapon in the world. To be sure, in recent years other inventors had come along and deftly updated the machine gun. Using a recoil system as propellant meant that a single trigger pull—rather than the turn of a handle-crank, as was the case with the original Gatling gun—now could initiate the firing. Soon to be developed would be the light machine gun and the submachine gun; the former would render machine guns portable enough for two-man crews to maneuver them around the battlefields of World War I a little more than a decade hence, while the latter would enable small-caliber ammunition to be used in weapons with continuous firing capacity. These changes would transform the machine guns of yesteryear—heavy, static things, things that looked more like the cousins of cannons than the brothers of rifles—into what we picture today when we hear the phrase "machine gun": lightweight, nimble weapons that pack astonishing rapidity and volume of fire into what is basically a handheld device.

Yet Gatling was the man who unlocked the secret of the machine gun. He had done it almost a half a century before this moment in 1903, even before the widespread use of metal bullets. Without the Gatling gun, subsequent weapons—the Maxim, the Gardner, the Lewis, and later in the twentieth century, the deadly bouquet of assorted assault rifles—could not have come along as quickly as they did. The Gatling gun was first. And the thing about which inventors, including Leonardo da Vinci, had dreamed for centuries—a multiple-firing gun that actually worked—suddenly was present in the world, created by a farmer's son with no formal training or engineering expertise, with only a supple mind and a driving will.

He was an old man now, and a weary one, and he had seen his gun outclassed and outsold by its rivals. Disillusioned by the armaments business, Gatling recently had switched his focus back to his first love: agriculture. The trip to St. Louis had been undertaken to establish the Gatling Motor Plow Company, which would manufacture the steam-driven plows sure to revive his fortunes.

But when his wife and his son suggested a rest, just a short break, the inventor had agreed. Richard and Jemima Gatling would stay with their daughter's family in New York and then, once he felt stronger, once he'd licked some lingering flu symptoms, he would head right back to St. Louis.

The editors at *Scientific American* welcomed him that morning. He had known them for decades. They had featured his gun on the magazine's cover many times, and reported on his many other inventions, too. In fact, the publication maintained its own patent agency and had handled Richard Gatling's patents for years. There could have been no better place, no more fitting locale, for Gatling to spend what would turn out to be the last few hours of his life.

He walked home, arriving at about 1:15 P.M. The trip had tuckered him out. He did not join his family for lunch; instead, he went into the study for a nap. A few minutes later, when the telephone in the study rang, Ida's young daughter, Peggy, came in to answer it. Her father, Hugh Pentecost, was calling from his office. While chatting, Peggy heard odd choking sounds from her supine grandfather. Her father advised her to call the doctor forthwith.

Dr. Charles P. Duffy, who lived just down the street, was summoned by messenger. One look at Richard Gatling, though, and he knew: This man was beyond the reach of medicine, beyond the help of any sort of earthly ministration. Gatling died a few minutes later, ringed by his wife, his daughter, and his granddaughter.

Given his stature at the time, it is not surprising that his death elicited long and respectful notices in major newspapers, including the *New York Times*—whose editors, some four decades earlier, reportedly had commandeered his guns to fight off an approaching mob. "Dr. Gatling Dead/Celebrated Inventor of the Gatling Gun" was the headline in the February 27, 1903, *Hartford Courant*. In an editorial published the same day, the paper proclaimed, "Any catalogue of the Hartford names that have gone round the world would include his. . . . The work that made him famous was wrought here; he remains for all time one of the town's assets."

Gatling's body was shipped back to Indianapolis by train, arriving at approximately 11:45 A.M. on March 2 at Union Station. It was taken to the home of John R. Wilson, a family friend. At the funeral later that day, the Reverend A. R. Benton made note of Gatling's strong Midwestern ties, and of the weapon that had made him famous: "For many years of his early life, his residence was here; here his family relations were made, and here his life's work was successfully begun. . . . Soon came the stress of our Civil War and he conceived the idea of mitigating the horrors of war by a more deadly weapon than any one then in use. Paradoxical as it may seem, it was his contention that the more destructive the weapons of war, the fewer are its casualties. In his earnest and convincing way he maintained his aim to invent such a weapon was humane, beneficent and philanthropic. Nor was his contention without reason, for he had all history behind him, supporting his claim. This destructive arm of warfare, now bearing his name, has become famous on two continents, and promises to perpetuate his name and fame to remote generations."

He was buried in Crown Hill Cemetery, surrounded by the somber gray obelisks marking the graves of many members of the Sanders family, and of the Wallace family, two of the great Indiana political dynasties of the early nineteenth century—powerful then, forgotten now. Similarly, Crown Hill Cemetery, first dedicated in 1864, is not the place it once was. It used to be on the outskirts of the city; a visit there meant a pleasant trip to the countryside. To a quiet, restful spot cushioned by trees and glades. No more. The city grew, as cities will, enveloping the 555 acres of Crown Hill. These days, the cemetery is surrounded by busy roads, by fast-food restaurants, by empty storefronts, by a crumbling and troubled neighborhood.

Gatling might not mind that. He appreciated change. His career, his life's work, was built around the idea of change. Like any good inventor, he depended on the endless human appetite for the new. If things stayed the same, if people were satisfied with doing what they'd always done even if someone came up with a better method, he'd have been out of a job. There would have been no need for a Richard Gatling. And no place for his imagination to rove.

There would have been no special meaning to the moment eighty-five years before when, in a house that stood up from the Carolina cotton fields, the third son of Mary and Jordan Gatling was born. He grew up with a keen eye and a sharp mind and indefatigable optimism. He fished and he swam in the Meherrin River, and he watched the boats dock at Murfreesboro, and he believed he could build a better propeller for those boats, a faster and more efficient one, and while he was at it, a better seed planter and a better cotton cultivator and one day, far down the road, a better gun. Growing up, he must have felt the future all the way down to his fingertips.

Gatling was gone, but his idea was indestructible. The immediate descendants of his gun would, in a few short years, transform war into something altogether different from what it had ever been before. And while the Gatling gun was declared obsolete by the U.S. Army in 1911, that was a bit premature. The Gatling gun was no relic. It would make a spirited comeback—and not simply as a cultural symbol. More than half a century after Gatling's death, General Electric would dig up and dust off his original 1862 patent to use as a template for the Vulcan gun, which became an iconic weapon in the Vietnam War. Dubbed "Puff the Magic Dragon," the Vulcan gun hung from the flanks of helicopter gunships, firing six thousand rounds a minute. It was used to extravagantly lethal effect in the war that, to many, still encapsulates the tangled conundrum at the heart of American history: how to be an empire without taking on an empire's arrogant and brutalizing ways.

It is almost as if the Gatling gun haunts the world, appearing at intervals to remind it of the dark side of technological endeavor, of the worst that can come from even the best of intentions. The

Gatling gun returns again and again, a ghost-gun, seeking reconciliation with the guileless imaginative forces that created it.

With the commencement of World War I in 1914, eleven years after Richard Gatling's death, his legacy became clear. The meaning of his mechanical genius and how it changed the world—not just its armaments, but also its attitude—emerged.

In the darkening days of August 1914, advisers to the king of Belgium—tiny, neutral Belgium, the small, quaint country which, by a geographical accident, was a stone in the boot of the advancing German army, a momentary annoyance that would have to be removed, perhaps forcibly, and forthwith—met to talk over the ominous signs.

Belgium's King Albert faced a wrenching dilemma: If he gave the Germans permission to march through his homeland on their way to France, it would render Belgium complicit in the attack and hence France's enemy. If he opposed the Germans, then Belgium would become Germany's enemy. The king and his advisers made their choice. Germany would be rebuffed, which meant, of course, that plucky Belgium would have to prepare for the fight of its young life as an independent nation—and the implication of that choice came down to this, according to contemporaneous notes made by Baron de Bassompierre, undersecretary of the Belgium foreign office: "If we are to be crushed, let us be crushed gloriously." Historian Barbara Tuchman offered a gloss on that moment: "In 1914, 'glory' was a word spoken without embarrassment, and honor a familiar concept that people believed in."

Honor and glory. Such words would be swept away in this war, along with eight and a half million lives and so much else, so many wrecked landscapes and torn-apart nations sent tumbling into long chaos by artillery and starvation. But the words matter, too. The abstractions. The revelation that concepts such as honor and glory were dead, as dead as a good chunk of Europe's young manhood. And had those concepts not been so persistently believed in, had they been recognized at the outset as empty words, the course of the conflict—one in which machine guns changed everything—might have been quite different.

It was, in Joseph Conrad's phrase, "the world's great storm." It was the war in which the mechanization of death found its first vivid display, when the full implications of what Richard Gatling had wrought became obvious. It was the war in which, as one historian noted, "scientists, engineers and mechanics would be as important as soldiers."

The nineteenth century did not end at the flick of a calendar page or the sweep of a clock hand. It was not a matter of chronology. The nineteenth century ended when British officers flung themselves into battle on horseback in 1914 and 1915, flourishing sabers over their heads, and were met with a synchronized storm of machine-gun fire from German troops. "When we started to fire we just had to load and reload," a German machine gunner later wrote. "They went down in their hundreds. We didn't have to aim, we just fired into them." The British had begun the war with some two hundred to three hundred machine guns on hand; the Germans had more than five thousand. The British, for a time, would continue to insist on the primacy of honor and glory; they would still wave sabers and declaim manly sentences about fortitude and fair play. So would their allies, the Russians. So would the French. All were dazed and caught and misled by what Tuchman called "the spell of élan." The nineteenth century ended when the last soldier raised a final sword against a machine gun during some otherwise unremarkable battle, after which honor and glory could no longer be spoken of, except ironically.

• • •

Sabers, though, were hard to leave behind. They were more than long metal objects; they were emblems of the way wars had always been waged. Valor and dash: Swords symbolized those attributes, and to abandon the weapon was also to let go of the noble dream, of the grand image of bold, telling strokes delivered during close combat. To give in to that reality was to eliminate much of what made war special—not to mention bearable. Swords were difficult to relinquish for the same reason that red pants continued to be worn by French troops in World War I: Yes, they made the soldiers far too vulnerable, standing out in a dun-colored battlefield like a summoning shout of "Here! Over here!" They made the French fighters easy targets. But trading those pants for darker, more practical garb meant acknowledging the war's insane savagery. It meant the loss of some final scrap of the beautiful illusion. Better an attractive pair of pants than an ugly truth. Better élan than survival.

Such stubborn nostalgia, ultimately fatal in its implications, crossed many borders. Sir Douglas Haig, who commanded the British Expeditionary Force, considered the machine gun "a much-overrated weapon," drastically inferior to soldiers charging on horseback. The old ways were hard to give up. The old ways had poetry in them, and romance, and dash. The old ways, however, were suicidal. "Our machine guns did excellent work," a German soldier wrote to his parents. "The English fell in heaps." It was the work of machines, and not the bravery of soldiers, by which war would henceforth be characterized.

General Sukhomlinov, the chubby and preoccupied Russian minister for war, reprimanded officers who taught at his country's war college for their interest in flashy new weapons such as machine guns and advanced artillery. Swords, he thundered, swords were still the answer. The bayonet charge still ruled. "As war was, so it has remained," he retorted, when brought news of some battlefield innovation. Even the Germans initially were obtuse. Sukhomlinov's German counterpart, Count Alfred von Schlieffen, wrote that certain things might have changed in warfare, but "the principles of strategy remain unchanged." They were old men in a very new kind of fight, and it would be a longer and ghastlier fight than anyone could have imagined. Technology would go to war, too, along with 65 million soldiers. Technology had changed the way people did everything—the way they traveled and communicated and conducted business. Yet as World War I began, traces of the old ways remained, too, producing odd combinations and strange hybrids of the new and the old that would have seemed amusing, had they not involved death, had entire countries not been at stake. Tiny and overmatched, the Belgian army, with no heavy field artillery, had less than half the number of machine guns per soldier as did the German army—but at least they *did* have some machine guns, yes? They, too, possessed technology? There was some cause for hope?

The Belgian machine guns were hauled by dogs.

In the first skirmish of the Spanish-American War a decade and a half before, Theodore Roosevelt had reluctantly given up his sword, finding that it slowed him down as he plunged and charged under furious fire from Spanish troops. "It kept getting between my legs as I went tearing through the jungle," he wrote. "I never wore it again in action." It was quickly replaced in his affections by Gatling guns. Others, though, could not make the switch quite so smoothly.

Wars had been fought one way for a very long time, and now they would be fought another way. When it came to changes in armaments over the centuries, there generally was a pattern: a prolonged period of stasis and then a sudden shift, a lurch, undertaken only reluctantly. For centuries, wars were personal. They were like private feuds writ large, replicated across a battlefield in a dark, rippling echo. Then they became epic sieges of mass death, of slaughter on an exquisitely grand scale, of bland and impersonal sweeps by technologically acute weaponry. Valor alone counted for little. The initial resistance to armaments such as machine guns by officers such as Haig and Sukhomlinov may be hard to fathom on an intellectual level, but on an emotional level, it is perfectly plausible. Machine guns "negated all the old human virtues—pluck, fortitude, patriotism, honor—and made them as nothing in the face of a deadly stream of bullets," one historian theorized. When machine guns were employed at sea, a disgruntled observer recalled, they seemed "sneaky and un-Nelsonian."

Almost unthinkable was the need to give up "the old notion of the glorious charge and the breakthrough, and relying upon the timely use of the bayonet, the sabre and the lance." British officers, reluctant to accept the new reality, went to war with their walking sticks. And then there was the French officer who, early in the war, responded to an enemy attack by mounting his horse and dashing into the battle while waving his sword over his head, having urged the regimental band to play "La Marseillaise." As comical as that sounds, it represented a way of thinking that was doomed—but before it went away, it was to have devastating consequences: The French, not understanding at first how machine guns would work in this war, maintained their blinkered belief in grand offensive strikes. French officers sent wave after wave of infantry forces against the German machine guns, with predictable results.

Artillery was crucial, too, of course, in giving World War I its unique horror, in making this war as long and as terrible as it was, but artillery was a familiar hell. Soldiers were accustomed to artillery. There had been heavy cannons and howitzers in battles for centuries. The great guns turned out by Krupp and other arms makers represented a difference in degree, but not in kind. Machine guns, though, instilled an entirely new class of terror. Poet Robert Graves, a World War I veteran, described the distinction between a shell from a cannon and a bullet from a machine gun: "A rifle bullet, even when fired blindly, always seemed purposely aimed. And whereas we could usually hear a shell approaching, and take some sort of cover, the rifle bullet gave no warning." Cannon fire seemed blundering; the cannon was a clumsy behemoth, striking all over the place, like a blind giant slapping at gnats, hoping for a lucky hit. Machine-gun fire was lithe and quick and seemed malevolent; in its blanketing hail of bullets, it felt cruelly inescapable, like fate.

Machine guns mattered. They mattered more than anyone quite realized until well into the war. The message, though, did eventually sink in. The British had started the war with skimpy stocks of machine guns but possessed the industrial capacity and technical skill to catch up: In 1911, Vickers Brothers produced twenty-six machine guns; in 1914 and 1915, the output was 377 and 2,433 machine guns, respectively; in 1918, the company turned out 41,699 machine guns. The British soon realized how important machine guns were, and, in the fall of 1915, organized the Machine Gun Corps. On the monument to the Machine Gun Corps in London's Hyde Park, a biblical verse—1 Samuel 18:7–8—adorns its pedestal: "Saul hath slain his thousands, but David his tens of thousands."

Blood and numbers were irrefutable proof of the superiority of multiple-firing weapons, but few really wanted to believe it. Few were prepared to accept that things had really changed that much, that warfare now was a question of machines and not will, not heroism, not honor. Machine guns worked well but felt wrong. "A weapon that could be used to kill soldiers impersonally, and at a distance of more than half a mile offended deep-seated notions of how a fighting man ought to behave," writes historian William McNeill. "Gunners attacking infantry at long range were safe from direct retaliation: risk ceased to be symmetrical in such a situation, and that seemed unjust. Skill of an obscure, mathematical, and technological kind threatened to make old-fashioned courage and muscular prowess useless. The definition of what it means to be a soldier was called into question by such a transformation. . . . [S]oldiers in general clung energetically to the old-fashioned muscular definition of battle."

Even the way the world had gotten itself into this war seemed different than in past conflicts. Even the run-up to war felt impersonal and machinelike, a matter of gears locking into place and pistons firing and springs tightening. There was a mechanical aspect to mobilization, a sense that once events had been set on their courses, there was no stopping them. No turning back. Only the great forward surge. Once initiated, the iron wheel turned steadily, engaging the sprockets, instigating motion all the way down the row. Germany mobilized more than three million men in less than a week. One thing led to another, and another, and then another. There was an assembly-line quality to this remorseless preparation for hell. "Reservists went to their designated depots, were issued uniforms, equipment and arms, forced into companies and companies into battalions, were joined by cavalry, cyclists, artillery, medical units," Tuchman wrote. "From the moment the order was given, everything was to move at fixed times according to a schedule precise down to the number of train axles that would pass over a given bridge over a given time."

It was "the blow that hurled the modern world on its course of self-destruction," Jacques Barzun said of World War I. It swept up thirty-two nations, leaving more than eight million dead and twenty-one million wounded. It scrambled the map of Europe and the Middle East. It was the logical consequence of the arms race that had developed along with the rest of the Industrial Revolution. If there was a better weapon out there, then a nation's leaders would be foolish and irresponsible not to want it, would they not? And when an even better weapon came along, then that weapon, too, would have to be procured, because one's foes would be aware of it, too. No one wanted to fall behind. And because of the rise of the great multinational arms firms, no one had to.

Arms races are a fact of global life. The first guns had been invented in about 1290, when the Chinese learned how to harness the propulsive powers of gunpowder. Yet the business of armaments grew and prospered much more rapidly in the West than in the East. The reason, William McNeill theorizes, can be summed up in a single word: capitalism. In the market-based economies of the West, there was ample incentive for craftsmen to continuously improve weapons for wealthy and impatient kings. The Chinese, while innovative in the laboratory, lacked the entrepreneurial urge that sparked so much Western achievement. Chinese culture in these early centuries taught that the concentration of too much money in too few private hands was immoral. Gunmakers in the West, bound by no such scruples, got down to work. The perfecting of deadly weapons held the promise of a big payday.

In the 1540s, European metalsmiths figured out how to cast iron in large enough chunks to make cannons. (Ironically, as McNeill notes, the technology was borrowed from the methods used to cast church bells.) Weapons artisans also realized that the secret was to make smaller, not bigger, siege guns; smaller guns could shoot denser cannonballs instead of heavy, cumbersome rocks. Those cannonballs could cause the same damage without the fuss and expense of hauling huge stones. Such weapons remained the standard for the next three centuries. In armaments, a great leap forward generally is followed by several hundred years of the status quo.

And so it was with the machine gun. Save for innovations such as the percussion cap and the rifling of barrels, the basic technology of firearms had remained relatively stable for many years. Then came Richard Gatling and his gun. In a swift flurry that seemed to mimic the actual firing of these new weapons, the market was overrun by hordes of machine guns. In the late nineteenth century, as the pace of the global arms race accelerated, as profits multiplied, the Gatling gun was overshadowed by such guns as the Browning, the Hotchkiss, the Lewis, the Vickers, the Nordenfelt, and, of course, the Maxim.

Hiram Maxim hated trees. That's how it seemed, in any case, given the number he destroyed while test-firing his machine gun. Trees were Maxim's targets of choice. When potential customers arrived at his rented London home in the early 1880s to size up the weapons that were quickly making a name for themselves, Maxim led his guests out to the five-acre field that surrounded the house, aimed the weapon, and let loose. Tree trunks exploded. Leaves disintegrated, bark was smashed to chips, limbs were obliterated. One summer, Maxim's potential customers included the Chinese ambassador to Great Britain and his entourage. A witness recalled that the "glades and lawns were decorated by groups of silk-robed figures" engaged in the sport of "cutting down trees by machine-gun fire."

Maxim was born in Maine in 1840 to an impoverished family, and his early life was similar to Gatling's. He had virtually no formal schooling. He was a self-taught mechanical engineer. And just as Gatling had gleaned the basic design for his first machine gun from an innocuous earlier invention—his seed planter—the Maxim gun likewise arose from Maxim's first foray into inventing. He had made a better mousetrap. The traditional mousetrap killed just one mouse at a time. Subsequent mice ate the cheese, which Maxim regarded as wasteful. So he designed a mousetrap in which the energy created by the trap's snapping on the first mouse would reset the trap, initiating a succession of mouse kills. Maxim credited the device with giving him the idea for the Maxim gun, the first of which he patented in 1883.

The Gatling gun employed a rotating barrel; the turning of the hand crank dropped each bullet into its chamber, whereupon it fired, just as the seed planter dropped seeds one by one into the furrow. The Maxim gun, though, used the power of the recoil from the exploding bullet—not the energy generated by a hand crank—to propel the subsequent bullet. Maxim's first guns could fire some six hundred rounds a minute. In 1884, he patented a machine gun that utilized the muzzle gases to propel the next rounds. Legendary gunsmith John Moses Browning would employ the muzzle-gas idea in his machine gun, unveiled in 1891 and dubbed the gas-hammer machine gun.

Maxim was just as enthusiastic a salesman as Richard Gatling. He was loquacious and flamboyant, jumping into contests with other types of machine guns and brashly guaranteeing victory. He had relocated to London early in his career and continued to improve his weapon there. The Maxim gun would become the Allies' signature machine gun in World War I.

Again like Gatling, Maxim never stopped inventing; Maxim called himself a "chronic inventor." In an eighteen-year period beginning in 1866, he took out eighty patents. He had actually begun working on gun designs in 1854, but had grown discouraged about the prospects for multiple-firing weapons, because the inadequate ammunition of the day—paper cartridges meant there could be no uniformity—seemed to doom attempts to create a gun that could discharge quickly and smoothly, without jamming. Maxim, in fact, took pains in later interviews to praise the Gatling gun, calling its workmanship and those of other early machine guns "exquisite." Maxim understood that, given the primitive state of ammunition in the 1850s and early 1860s, it was astonishing that Gatling guns fired as well as they did. It was a tribute to Richard Gatling's technical prowess.

Maxim made his living in the 1860s and 1870s by becoming an expert in the installation of gas lighting, but he kept an eye on the arms trade. Later, he would famously quote a friend who persuaded him to get back into the gun business: "Hang your chemistry and electricity! If you want to make a pile of money, invent something that will enable [them] to cut each others' throats with greater facility." Harnessing the recoil energy, the single barrel of the Maxim gun—fed by a belt of ammunition—could fire six hundred rounds a minute, all at the touch of a finger on the trigger.

Other machine guns, too, had developed in the wake of the Gatling gun. Some of these early machines looked like Gatling gun knockoffs: Benjamin Hotchkiss's gun, patented in 1871, had a hand crank like the Gatling gun. Yet the Hotchkiss operated more like an artillery piece, with larger-caliber ammunition and a fuse-lit firing system. The Gardner gun, invented in 1874 by William Gardner, employed a feed system similar to that which Gatling added to his guns in the 1870s. Arms makers in Sweden and France also produced machine guns: A Swedish businessman named Thorsten Nordenfeldt gave his name to the Nordenfelt gun, which used a gravity feed system like the Gatling gun; Great Britain bought a great many Nordenfelts in the late 1870s, infuriating Richard Gatling, who felt his design had been copied. And France's Montigny mitrailleuse, twenty-five barrels strong, reportedly achieved a firing rate of 250 rounds per minute.

But these machine guns all were mechanical. They were powered by hand cranks or by levers. It was the Maxim gun and its recoil system, a system shared by the British-made Vickers and the Austrian-made Schwarzlose and many others, and then the gas-operated machine guns such as the Lewis, which upended the armaments world and, by many assessments, turned World War I into the extended inferno it was.

• • •

In 1877, Richard Gating had been asked by a young correspondent why he invented such a destructive device. "It occurred to me," he wrote back to her, "that if I could invent a machine—a gun—which could, by rapidity of fire, enable one man to do as much battle duty as a hundred, that it would, to a great extent, supersede the necessity of large armies. . . ." It sounds like a rationalization for having bestowed upon the world a very bad thing, but actually, Gatling has been proven right—and here, the analogy between the Gatling gun and the atomic bomb once again is relevant, beyond the fact that the creators of both lethal technologies had only the best and most humane of motivations. During World War II, the United States and Great Britain began to "industrialize war—shifting resources into artillery, tanks, warships, and, above all, aircraft," Niall Ferguson points out. "In effect, investment in bombers reduced casualties among Allied servicemen and greatly increased casualties among Axis civilians, a process that culminated at Hiroshima. Once dominance of the skies had been established, ground forces could be used at a far lower cost to life and limb." As Gatling had predicted, the more deadly and effective the technology used in a war, the fewer the numbers of human beings required to fight it.

Technology also made war cheaper, which was always a central part of Gatling's sales pitch: Buying my guns, he maintained, is actually thriftier than not buying them. Civil War generals did not purchase Gatling guns in sufficient quantities to test this theory. But in the twentieth and twenty-first centuries, with the spectacular development of armaments technology—technology that would, one might think, be so exorbitantly expensive as to strain any nation's budget—Gatling's point is made. "Technology has lowered rather than raised the costs of war," Ferguson writes. "In terms of 'bangs per buck'—destructive capability in relation to expenditure—military technology has never been cheaper." The same mass-production techniques introduced by the Industrial Revolution, the ones that gave people cheaper shovels and toasters and earmuffs, also produced cheaper weapons.

And there was another part of Gatling's argument, too, that the world would have to live its way into understanding: the concept of a weapon as deterrent. John H. Parker, whose expertise with Gatling guns in the Spanish-American War earned him the respect of Theodore Roosevelt and the sobriquet "Machine Gun Parker," believed that machine guns just might do their best work before, not during, the moment when they are unleashed on the enemy. "The destructive effect of machine guns coolly handled at close range is frightful," Parker wrote in one of his machine-gun handbooks. "The moral effect is even greater. When the enemy finds that every man who exposes his head above the trench to fire receives a bullet in the head, it produces a panic." You can almost hear the ghost of Richard Gatling, murmuring his approval and thanks. In Gatling's 1863 sales brochure, one of the first for his new gun, the inventor wrote: "It is confidently believed that no body of troops could be made to withstand the fire of such a death-dealing weapon, for the reason that men will not fight on such terms of inequality, or when there is no chance of victory." Gatling had predicted what Parker had witnessed: Machine guns can be as effective when quiet—but ominously present—as when they are activated. What they do is deadly enough, to be sure, but it is not what they do that defines them. It is what they *might* do, in the next few seconds.

CHAPTER EIGHT

WARRIORS AND SAGES

A multitude of simultaneous deaths appears to us exceedingly terrible.

—Sigmund Freud

A few Sokolov Maxim 7.62-millimeter machine-guns, which resembled farm machinery with their two wheels and towing yoke, their fat barrels pointing backwards as if to drop leaden seeds into the fields (five hundred per minute of them)—how ludicrous!

—William Vollmann

Among the souvenirs that visitors to the 1876 Centennial Exposition in Philadelphia might have toted home was a small, rectangular advertising card. The cards were handed out at the U.S. Army pavilion during the six-month event. On one side was an elegant drawing of a Gatling gun; on the other, a courtly but forceful reminder of the glorious weapon that the fair's approximately ten million attendees could have beheld, were they so inclined, alongside other modern marvels such as typewriters and telephones and sewing machines and bottles of Heinz ketchup and an odd new snack consisting of popped corn:

> As a practical military machine gun the GATLING has no equal. It fires from 800 to 1,000 shots per minute, has great accuracy, and the larger calibres have an effective range of over

two miles. The following calibres are made: .42, .43, .45, .50, .55, .65, .75, and one inch. It has been adopted by nearly all the principal governments of the world.

Gatling Gun Company,
Hartford, Conn., U.S.A.

The Gatling gun took home a souvenir of its own from the exposition: the medal for best machine gun, beating out rivals such as the Gardner and the Hotchkiss. Richard Gatling, who had attended the fair at least twice, bringing along his wife and his daughter, was pleased; he had lobbied hard for the honor. "Do not forget to suggest to . . . the Judges, that the Hotchkiss gun in most of its main features is copied after the Gatling gun," Gatling wrote to his sales associate, Edgar T. Welles, two weeks after the exposition opened. "The Judges should not give an imitator, or a copyist, an award over the original invention: The Gatling however is a better gun than the Hotchkiss & fires with five times—yes, ten times the rapidity." Gatling's promotional push paid off. On September 27, 1876, his gun received the bronze medallion, four inches wide and absolutely invaluable, Gatling believed, for future bragging rights.

In 1876 the Gatling gun was regarded as a laudable American accomplishment, another example of native ingenuity and craftsmanship and problem-solving acumen: America at its muscular, can-do best. And the gun's inventor, often referred to with more courtesy than accuracy as "Dr." Gatling, was depicted in formal portraits as a solid, staid, down-to-earth businessman, decked out with the requisite snowy beard, serious expression, close-fitting vest, and well-tailored coat, with a watch fob draped importantly across the portion of that vest revealed by the unbuttoned coat, as if tethering him, body as well as soul, to the ceaseless sweep and manic pull of time, to the chronological imperatives of American business, the constant forward march.

Stories about Gatling drove that image home. He appeared in top magazines of the day, from *Scientific American*, which often featured details about the latest improvements to his famous gun as well as his many other inventions, to mass-circulation, general-interest publications that the whole family might read and savor, such as *Harper's Weekly* and *Potter's American Monthly*. These magazines were part of a powerful new force in American culture that was, historian Richard Slotkin states, "in the process of rapid development and expansion of resources, productivity and market reach" throughout the nineteenth century: the mass media. Gatling was a celebrity inventor in the league of a Thomas Edison or an Alexander Graham Bell. But he was also a celebrity businessman, such as a John D. Rockefeller or an Andrew Carnegie. Gatling was both inventor and tycoon. He was a new kind of American hero: the self-taught mechanical whiz who worked with his sleeves rolled up and his tie unstrung; and the successful businessman who moved with ease in a pressed suit and starched collar.

Then came the twentieth century, and the systematic catastrophes enabled by ever-more-technologically sophisticated weaponry. Machine guns turned World War I into a defensive stalemate, stretching the war out across years of futile, ghastly combat. Later in the century, machine guns thrust phenomenal killing power into the hands of guerrillas and insurgents in hundreds of small but excruciatingly bloody conflicts. And Gatling, a man who had once been admired as a pioneer of technology, a visionary, an archetypal American success story, dropped out of the national story altogether. What happened?

Admired he surely was, and famous, too, as famous as any movie star of today, any musician. People were captivated by Gatling. So it was no wonder that in 1879, the editors of *Potter's American Monthly* sent a writer to Hertford County, North Carolina, to collect details about the early life of Richard Gatling, genius gunmaker.

Gatling didn't live there anymore—he hadn't lived there for more than three decades—and naturally the writer, Charles H. Foster, knew that. But because this was the place where Gatling came from, this was the place where Foster came. He came to get a sense of the great man—the man whom Foster will describe, in the piece he writes for the May 1879 issue, as "one of the foremost and conspicuous of living American inventors, and one whose fame is world-wide, and whom barbaric as well as Christian nations have joined in honoring." In the emerging genre known as the magazine profile aimed at a broad audience, detailed descriptions of the famous person's origins are practically compulsory. Origins help knit legends, and legends boost sales. Gatling apparently did not accompany Foster on this visit, but he surely approved of it; by this time, Gatling knew a little something about public relations.

Gatling's name was familiar to a vast legion of people who had never met him—a situation that had been possible in earlier eras only for kings and gods, and an important milestone in the history of fame, brought about by mass circulation publications and the breakthroughs in nineteenth-century communication technology. Because of the explosive population growth in the United States throughout the 1800s, individuals had begun to matter far less; the sheer numbers now thronging the land guaranteed it. Yet at the same time, the potential routes to personal fame were growing, too: There were more newspapers, more magazines, and there was the rapid development of a fantastic new image-preserving technique known as photography. "As every advance in knowledge and every expansion of the world population seems to underline the insignificance of the individual," writes historian Leo Braudy, "the ways to achieving personal recognition have grown correspondingly more numerous." The blur of the mass made the rise of a single person to fame—a Richard Gatling, a Thomas Edison—seem all the more unlikely and remarkable. Yet every famous person answers some need in the culture at large, Braudy says; the famous are "vehicles of cultural memory and cohesion." Fame is the place "where personal psychology, social context, and historical tradition meet." It is important to understand the scope of Gatling's fame in the late nineteenth century and the probable reasons for its pervasiveness. Otherwise, Gatling's slide from notoriety to obscurity might be dismissed as just another cruel trick of fate; unfortunate, perhaps, but not especially illuminating.

The life stories of famous people—how they got where they were, preferably by overcoming some colorful hardships along the way—often were packaged by nineteenth-century writers as inspirational parables, as cheerful and uplifting morality tales. By the mid-1800s, British author Samuel Smiles had ritualized the technique with his biographies of great engineers. The message was not in the least subtle: Why, with a little pluck and a lot of confidence, why, you—yes, *even you!*—can reach the top. His books carried titles such as *Self-Help* (1859) and *Thrift* (1875). It was the essence of nineteenth-century earnestness. Success could be achieved by emulation. And hard work, too, of course.

So when Foster shows up at what he calls "the old Gatling homestead," he knows which details to jot down, which details will tickle his readers and fulfill their expectations for didactic uplift. He knows what they'll want to know about the formative years of the famous inventor. He walks up the wide, neat avenue connecting house to road. "Not a leaf, or straw, or scrap of paper, or a particle of any other sort of litter is to be seen anywhere in the whole large yard," Foster later will write in his article, titled "The Modern Vulcan." He makes a catalog of the trees: "spruces, cedars, cypresses and junipers, along with the elm, maple and ash." He tastes the apple cider offered by James Henry Gatling, Richard Gatling's bachelor brother, the one who stayed on the farm while the other brothers left to seek their fortunes. Foster is especially intrigued with an old cabin next to the main house. He will strive to infuse the spot with a homespun Lincolnesque veneer: "It is the original family residence," Foster writes, "and under its humble roof was born the great inventor who now visits as a welcome guest the stately halls of kings and emperors."

In his published story, Foster chronicles the journey of Richard Gatling's father, Jordan Gatling, "a poor boy," who scrimped and saved and denied himself until he was able to buy eighty acres of distinctly unpromising land near the Meherrin River. "To this small beginning," Foster writes approvingly, "he added gradually, until, by patient economy and industry, he had acquired in one body twelve hundred acres of excellent land, a goodly part of which is in timber of highly prized long-leafed pine." Jordan's wife, the inventor's late mother, was "an angel of mercy and relief to the entire neighborhood." Such quaint details are part of the quintessential American success story, details with which Foster's readers are warmly familiar: the quiet lives of humble people who ask for no special favors, just the chance to work hard and prove themselves.

In the main house, Foster is led about by James Henry Gatling, described as "a prosperous farmer . . . without a dollar of indebtedness." James Henry Gatling has turned the house into a sort of Gatling family museum, and Foster is awed. He sees engravings of the Gatling gun, framed and hung on prominent spots on the parlor wall. Foster sees award certificates that Richard Gatling received for the agricultural implements he invented before the gun. Foster marvels at "the oldest piano in Hertford County." He exclaims appreciatively at the collection of walking sticks carved by Jordan Gatling shortly before his death. The sticks are "covered with miniature serpents and other reptiles, fishes, raccoons, etc., all perfectly life-like in color, shape, proportion," including one that overwhelms him with its delicate artistry. It is "completely covered with poems, essays, names of philosophers, poets, warriors and sages, and contains 510 words of 3,693 letters, all in clearly cut Roman text." Foster also gets a look at Richard Gatling's first invention, now "rusty with age": the screw propeller. This, Foster tells his readers, is the device that, although he was unable to patent it—John Ericcsson beat him to the punch—started Richard Gatling on the path to fame and riches.

The story's conclusion rises to a crescendo of impassioned inspiration, as Foster confidently places Richard Gatling in the great pantheon of "armorer[s] of nations": "We have seen him pass from a log cabin in the Carolina pine woods, on and up, until he enters imperial palaces, unchallenged. We have seen him, the new Vulcan, giving to the world's armies a weapon more potent than the bolt of Zeus. . . . But the story of Richard Jordan Gatling is more than this. In him is illustrated the true glory of our American institutions in the opportunity they offer the humblest lad to conquer, by honest and heroic labor, all obstacles of circumstance, and to achieve for himself a manhood which shall make him the acknowledged peer of princes."

How, then, did Richard Gatling go from being "the acknowledged peer of princes" to a forgotten man? What caused his name to disappear from history, even as the name of his invention lives on as, if nothing else, a cliché of hectic, percussive dread? And how did guns themselves, for that matter, go from being treated as practical necessities, fully worthy of display at trade fairs touting national progress, to objects of sinister ill repute?

Part of what happened is that the United States slowly backed away from the nineteenth-century conviction that armaments are crucial to the national identity, that well-intentioned countries can and should spread themselves and their way of life around the world. Cynicism replaced earnestness. And a thesis that once had seemed completely reasonable—that a gunmaker could be a purveyor of peace—began to sound instead like coy opportunism, if not rank hypocrisy.

There was a time, early on, when the United States really did believe that it had a special role to play in the world. There was a time, that is, when the word "empire" was not pejorative. The United States had embraced the idea of empire from its very birth, as historian Niall Ferguson recounts in his book *Colossus: The Rise and Fall of the American Empire.* There "were no more self-confident imperialists than the Founding Fathers themselves." Initially, George Washington referred to the newly formed United States as a "nascent empire" and "infant empire." Thomas Jefferson, Alexander Hamilton, and James Madison all wrote approvingly of the notion of a country that could and would expand at will. Throughout the nineteenth century, first by rolling out across the continent in a steady series of land purchases and war-won seizures of land, and later by reaching out for territories off the mainland, the United States grew. And grew still more. In 1893, Queen Liliuokalani was overthrown and the eight-island chain known as Hawaii began its journey toward eventual statehood. The major weapon employed to cow the queen's supporters: a Gatling gun.

In the 1800s, Americans thought imperialism was a fine idea. By the late 1900s, they no longer thought it so fine. And when a new ambivalence about arms and armorers set in, a name such as Richard Jordan Gatling's began to lose its luster, its prominence, and soon faded away. The notion that Gatling might have been sincere in his assertion that he had invented his gun to save lives, not cost them, went from plausible to laughable. His attempts to turn a profit from firearms made him seem not admirable as an entrepreneur, but despicable as, in effect, a war criminal.

The fact that arms are necessary to a nation's survival is a grubby and uncomfortable truth. Yet a vital truth it is, as historians long have recognized. Without superiority of arms, the Western world would have fared quite differently in its competition with other regions. "The Industrial Revolution in Europe and America," writes Geoffrey Parker, "forged some new tools of empire—such as the armored steamship, the rapid-fire gun." From the sixteenth through the eighteenth centuries, Parker continues, Islamic states expanded and solidified their grip on the world. The Ottoman empire dominated more than a million square miles of land, "which stretched from Morocco through Egypt and Iraq, to the Balkans and Hungary. So many states and societies were overwhelmed." What turned the tide for the West, Parker says, was innovation in arms and arms production. "Once the Industrial Revolution began to transform the production of gunpowder weapons systems in Europe, the Muslim states' lack of bureaucratic and financial institutions necessary to support capital-intensive and constantly changing military establishments, by land and sea, became critical. . . . Only military resistance and technological innovation—especially the capital ship, infantry firepower and the artillery fortress: the three vital components of the military revolution of the sixteenth century—allowed the West to make the most of its smaller resources in order to resist and, eventually, to expand to global dominance."

It is difficult to overstate the significance of arms in the success of particular cultures, as Paul Kennedy notes in his book, *The Rise and Fall of the Great Powers*. Military might is how the West won: "The improvements in the muzzle-loading gun (percussion cap, rifling, etc.) were ominous enough; the coming of the breechloader, vastly increasing the rate of fire, was an even greater advance, and the Gatling Guns, Maxims, and light field artillery put the final touches to a 'firepower revolution.'" Kennedy dubs this the "firepower gap," and gives it credit for "the global dominance of the West." Jared Diamond, in his popular and influential book *Guns, Germs, and Steel: The Fate of Human Societies*, would concur. But he traces this truth—that weapons have had a crucial impact on the success and longevity of various civilizations—back even further in human history: "Literate societies with metal tools have conquered or exterminated other societies. . . . The whole modern world has been shaped by lopsided outcomes," he writes, adding, "Technology, in the form of weapons and transport, provides the direct means by which certain peoples have expanded their realms and conquered other peoples. That makes it the leading cause of history's broadest pattern."

Yet Diamond is skeptical about the "great man" theory of history, so popular in the nineteenth century. Instead, Diamond chalks up the disparity in arms development across cultures to factors such as population density; political organization and whether it encourages or stymies food distribution; the hardiness of the environment and, within that environment, the presence or absence of wood, metal, and other weapons-making materials. To Diamond, the determining element in a society's rise to military power is not, after all, the energy and vision of a Richard Gatling, but elements such as soil composition, rainfall totals, and bureaucratic structure.

Improvements in arms may come down to the Gatling factor—to, that is, the genius of a key individual—or they may come down to what Diamond calls "several constellations of environmental factors." But the reality remains: Guns matter. Nations must constantly expand and improve their armaments if they intend to prosper—or even to survive. And these are facts that tend to make people shudder. They are facts from which some Americans in the late twenteith and early twenty-first centuries increasingly have wanted to turn away. The United States, despite its great success in the wars of the twentieth century, began to be "at most, a reluctant imperialist," writes historian John Gordon Steele, noting that America was the only major power that did not expand its territory in the victorious wake of World War I.

Thus the sweet earnestness of the nineteenth century ran headlong into the sour cynicism of the late twentieth and twenty-first.

That cynicism is justified, to be sure, given the horrors of terrorism and world wars, but it tends to make the optimism of the nineteenth century look almost ridiculous in retrospect, almost childishly naïve, and to make those nineteenth-century dreamers look foolishly gullible. To think that financial giants such as John D. Rockefeller might have been sincere in their religious convictions—as they were destroying business rivals and strong-arming public officials to craft legislation advantageous to their cause—is a leap of credulity many people are unable to make in the modern age. Yet the destruction and the strong-arming were, these titans believed, part of the competitive nature of business. They offered no apologies because to them there was nothing to apologize for. And once they secured their fortunes, their burden grew larger, not smaller. Rockefeller, among others, was demonstrably serious about enlightened capitalism, about the immense social responsibilities of immensely wealthy individuals. He established colleges. He gave a significant portion of his fortune to charity. Likewise, J. P. Morgan was almost singlehandedly responsible for the development of the Metropolitan Museum of Art, donating money and objects from his own collection. You can look at this largesse and see the work of guilty consciences—or you can look at it and see a time in American history when great fortunes automatically obligated those who amassed them to be generous, civic-minded, and farsighted. Rockefeller's son, John D. Rockefeller, Jr., as equally reviled as his father by some, taught Sunday school for many years. He tried valiantly to square up the contradictions. "My problem," he wrote, "was to reconcile right and conscience with the hard realities on a practical level." Ron Chernow, biographer of Rockefeller Sr., realizes modern readers may doubt that billionaires can possess moral scruples. Hence Chernow must explain that "Rockefeller's life was of a piece and . . . the pious, Bible-thumping Rockefeller wasn't just a façade for the corporate pirate. The religious and acquisitive sides of his nature were intimately related." A nineteenth-century observer might very well have accepted that at face value. A twenty-first-century observer needs to be convinced.

Mass market. Mass media. Mass appeal. "Mass" was a new adjective for a new era. A mass market enabled the rise of superstars such as Buffalo Bill and Annie Oakley, and Richard Gatling and Thomas Edison, too. It created a unified aggregate in America, one that could be shaped, swayed, mobilized, theorized about, manipulated, entertained, sold to. "The development of a true *mass* media with a national market begins in the 1830s and 1840s; but it is not complete until the 1890s," writes Richard Slotkin. Harnessing the newly awakened power of the newly developed mass audience, in fact, was a critical aspect of Edison's career as an inventor, according to historian Randall Stross. By pioneering "the application of celebrity to business," Edison "invented the modern world," Stross maintains. Like Gatling, Edison and his latest breakthroughs frequently were the subject of articles in *Scientific American* in the 1870s and 1880s. Adoring newspaper stories and fawning magazine profiles made both men household names, further sharpening the public appetite for information about them and their doings and more firmly establishing the revolutionary power of a transformative new entity known as a mass market.

Until the 1880s, a mass consumer culture in the United States largely did not exist. There were no department stores, and no such thing as pitching one's wares to a single large market. But with the rise of the steamboats, and then the railroads, with the quickening dissemination of information through the telegraph and through magazines with national circulation, a new idea was born in the late nineteenth century: the mass market. Department stores—Stewart's in New York, Wanamaker's in Philadelphia, Marshall Field's in

Chicago—opened in large cities. Public tastes and desires were nationalized, thanks to the advertising in those newly created mass-market magazines such as the *Saturday Evening Post* and the *Ladies' Home Journal*.

Raw numbers mustered their own logic and legitimacy. And just as the Gatling gun was demonstrating—chillingly, but irrefutably—the value of economy of scale when it came to what weapons could do, the consumer culture of the United States was demonstrating what that same idea of scale could do when it came to selling products and services. Indeed, throughout the nineteenth century, so many attributes of the modern world began to reveal their dual natures, their sunny upsides and their baleful downsides. Steamboats enhanced transportation—and brought about smallpox epidemics. The Industrial Revolution meant that all manner of goods now could be made faster and cheaper, and it created new jobs—and it also wreathed cities in choking pollution and kept communities at a constant boil with vicious disputes between laborers and bosses. Mass markets had a positive side, and a negative one. The 1880s and 1890s saw "near-volcanic change," writes historian Alan Trachtenberg. More people began to work for corporations, rather than for themselves, further blurring their distinctiveness. "The axis on which society turned," he adds, "was the shift from one form of capitalism to another, from predominantly self-employed proprietors to large corporations run by salaried managers." Gradually relinquished in the early twentieth century would be a certain poetic individualism, the idea of the lone genius creating the future with her or his indomitable vigor. The world was becoming a safer place, perhaps, but definitely a less adventuresome one.

The nineteenth century is the crucible of the modern world. Not only in terms of the gadgets and innovations, the steamboats and railroads, the safety pins and zippers and ice cream cones and machine guns, but also in terms of cultural ideas. The ancient, fate-driven world of castes and iron inevitability had given way to the great age of earnest self-direction. To self-made fortunes. To a belief in the golden possibility of change and regeneration. It was "the age of energy," as Howard Mumford Jones termed it, and the United States was the most energetic place of all; the nation in the nineteenth century was "heady, adolescent, picturesque, unpredictable, yearning to be loved and fearing to be outdone." America, in fact, seemed to be the validation of the most prominent philosophical ideas of the age. Thinkers such as Thomas Carlyle and G.W.F. Hegel held that history was "the march of progress from lower to higher forms," as Paul Johnson writes. History was "a positive force with an irresistible momentum." Herbert Spencer's ideas about the compulsive forward momentum of evolutionary progress entranced capitalists such as Andrew Carnegie. Major minds of the era believed that fates could be shaped and hammered according to human desire, like hunks of hot iron on a blacksmith's forge.

But as bright and glorious as all of that sounded in the nineteenth century, by the middle of the twentieth, it was dragging a shadow. Two world wars had put the lie to the idea that human beings were motivated by much more than greed and selfishness and bloodlust. Religion, rather than a force for enlightenment, had been unveiled once again—just in case humankind had forgotten, in the midst of all that nineteenth-century earnestness and optimism—as a force that also could ignite chaos and death. "History," William James reminded the world just after the twentieth century began, "is a blood bath."

And as that world began to come to terms with the reality of weapons capable of causing destruction on a vast, impersonal scale, the movement toward mass markets began to revolutionize American consumer culture. By the late nineteenth century, an intimate world of first-person encounters was vanishing. It was vanishing along with the country store at the crossroads, as more and more people moved away from farms and into cities, and it was vanishing from the battlefield, where technology was rendering individual valor largely irrelevant. The concept of a mass of potential customers—so alluring to merchants—was, to arms makers, equally alluring as a mass of potential targets. And while every era has its good sides and its bad, its advances and its retreats, there is something seminal about the last half of the nineteenth century, something momentous and definitive. The very best that the human mind could achieve—improvements and discoveries in science and manufacturing that lengthened the life span and enabled the leisure from which could arise magnificent literature and music and visual arts—also created the means to achieve the very worst: wars, genocides, holocausts. The Gatling gun is among the most compelling embodiments of this fraught paradox: a mechanical marvel that revolutionized the world's capacity for cruelty. Its inventor was a creative destroyer.

The Gatling gun, then, is an important cultural symbol, and continues to be so into the twenty-first century. But it is not just a symbol. Richard Gatling's inventive genius still influences armaments development. In February 2006, officials at the Lawrence Livermore National Laboratory in Livermore, California, at which nuclear materials are stored, began arming their security staff with modern, six-barrel Gatling guns known as Dillon Aero M134Ds, capable of firing fifty rounds per second. As the director of the National Nuclear Security Administration announced, "What we want to do is equip our protective force with the capability that will leave no doubt about the outcome." More than a century after Richard Gatling's death, the Gatling gun is still on the job.

Like the writer for *Potter's American Monthly,* you go to look for Richard Jordan Gatling in a place where you know you won't find him: the Gatling plantation, or what's left of it. You travel to an area on the outskirts of Murfreesboro, North Carolina. You step carefully through a cotton field, watching out of the corner of your eye for the wily snakes that are said to lurk here, until you come to the old cemetery. Passing through a gap in the wrought iron gate, you may be aware of that odd sensation many people get in cemeteries: the feeling that you've been expected for quite some time.

The headstones are clustered in unbalanced rows in front of a tall thin obelisk. *Erected A.D. 1860* is the inscription at the base of the white Italian marble structure, an inscription still clearly visible today. Carefully and meticulously carved on each of the four sides are brief biographies of the Gatling clan, the etched and necessarily truncated histories of the people who lie beneath those small headstones, the record of the family that once owned these vast fields that seem to stretch in all directions, right to the edge of the world: *Jordan Gatling/Born May 14, 1783/Died April 13, 1849/Aged 64 years & 10 months & 29 days / He was a kind father.*

And this: *Mary, wife of Jordan Gatling/Born Oct. 30, 1795/Died Sept. 30, 1868/A devoted Wife and affectionate Mother and a true Christian.*

On it goes, all around the obelisk, each side of the weather-stained sentinel yielding up its snippet of information about a particular Gatling, about Jordan and Mary and their children. The cemetery is a few miles outside Murfreesboro, a small town in northeastern North Carolina that, except for the cars and the telephone poles, appears little changed these days from what it must have been like a century and a half ago. The sun still presses its palm down on the world with an intensity that seems almost personal. A lot of the people here still make their living from the soil, getting up early each day to tend to peanut and cotton and watermelon crops.

The Gatling family cemetery used to be securely bordered by that waist-high, black wrought-iron fence, an intricate and beautiful thing with a hinged gate across the top of which *JAS. H. GATLING*—for

James Henry Gatling, Richard's brother, the oldest son of Jordan and Mary—is spelled out in an oddly delicate-looking iron swirl. But the fence of late has had a hard time of it. Sections have broken off and toppled. The gate itself is propped against a holly tree that rises next to the obelisk, matching its height, as if both are secretly standing on tiptoe to see who's taller.

The family home that ought to be just a few hundred yards away was demolished in 1979 by one of the dozens of owners who have held title to these grounds since 1881, when Richard Gatling, with no plans ever to live here again, began selling it off piecemeal. Cotton fields now surround the cemetery, row upon row upon row of the petite green plants whose leaves conclude in a funny little ball. Within that ball lies the fuzzy white stuff constituting the treasure and the heartbreak of the South, its gift and its curse.

The last of these Gatlings died a long time ago, but the cemetery is still here. Even in its dilapidated state, it is suffused with dignity and decorum. There's a quietness here. There's a contemplative sort of calm here, too, and a rich, almost voluptuous sadness. There's a sense of the past here, and of its everlasting mystery. One thing, however, is not here: a marker for Jordan and Mary's third son, Richard. The man who made the Gatling name internationally famous when he placed it on the world's first working machine gun is buried not here in the South, next to his parents and brothers and sisters and across the way from his grandparents. Not here amid the fields and woods where he was born in 1818, not here where he worked and played.

He is buried in Crown Hill Cemetery in Indianapolis. He is buried in the North, in the region against which his family and the close acquaintances of his childhood arose in such furious, ultimately futile opposition. Thus of all the contradictions in Richard Gatling's long and complex life—he was a peace-loving arms merchant, an inventor who dreamed of rendering war obsolete by making it more terrible still—the oddest and most perplexing conundrum may come down to this: Gatling's final resting place.

So there is no marker for Richard Gatling here amid his kinfolk's headstones in North Carolina, a place visited regularly nowadays only by dragonflies. It is true that he left the farm in 1844, when he was twenty-six, bound for St. Louis, determined to make his fortune on the strength of his ideas and the extent of his pluck. Yet he came back many times over the years, back from the great world beyond the Meherrin River. Sometimes it was to settle family business, but other times, it was just to say hello. He never lost touch with the people here. But neither did he choose to spend his eternity among them.

His country, too, has been beguiled by violent and obvious contradictions. It, too, has exhibited cruelties that smack not of calculation but of a ghastly moral myopia, of disparate layers of desire and distorted self-image. His country claimed that its government was the servant of the people, not the other way around, but then used government forces armed with Gatling guns against striking workers in the late nineteenth and early twentieth centuries. His country brayed importantly about freedom for all—but for more years than many care to remember, condoned slavery. Gatling and his parents and his siblings and his friends and his in-laws all owned slaves.

The slaves from the Gatling plantation reportedly are buried just outside the wrought-iron parameters of the Gatling family cemetery, anonymous in death as they were in life—except, of course, to their creator, who presumably requires no headstones to sort out who's who.

We know him, but we don't know him. We know his gun—the one invention of his that lives on as cultural emblem as well as in tangible form, the one among so many of the things he created—but we don't know him, not really, because history is inherently unfair, and sometimes downright perverse. History doesn't always reward the deserving with long fame, and occasionally seems to delight in elevating to immortality those of lesser achievement. And while the words "Gatling gun" are commonplace and familiar, while they constitute a household phrase, a metaphor for any rapid-firing spew of anything, the man responsible for it is largely forgotten. He slides into the white historical mist like a flap tucked into an envelope.

The fact that Richard Gatling is little known today constitutes a reminder, perhaps, of Americans' continuing discomfort with the implications of awesome military power, even when wielded with the best of intentions. Americans have yet to assimilate—and may never want to assimilate—the contradictions inherent in what Gatling was, and did, and how his great invention was put to use by a world that spent much of the twentieth century with its dukes up and its teeth bared.

In Richard Gatling's lifetime, the world became modern. He was born in 1818, when people traveled on foot and on horseback, and later by steamboat and railroad. Yet the year he died—1903—was the same year the Wright Brothers achieved the first powered aircraft flights at Kitty Hawk, North Carolina. From dirt roads and rivers to clouds and stars: Gatling's lifespan encompasses what must surely be the most amazing period of change in world history. It also, though, encompasses one of the bloodiest. Perhaps the perverse contrast is not quite the coincidence we might wish it to be.

Thus we come to know not only Richard Gatling but also the times in which he lived, the astonishing and improbable nineteenth century, at once so far away and so close. So distant and, mysteriously, so present, too, as we look around and see where it's left us.

When he set out from this North Carolina cotton field in 1844 to make his way across the tree-walled, river-floored wilderness of a new world, the woods didn't part for Richard Gatling. He parted the woods. And that, as each of us instinctively knows, makes all the difference.

ACKNOWLEDGMENTS

We dwell, all the livelong day, at the center of concentric rings: family; friends; colleagues. My mother, Patricia Weed, first introduced me to the insinuating mystery of the written word. Great thanks go out to her and to her husband, Don Weed. My older sister, Catherine Mary Dougherty, is the wittiest person I know, and a discerning reader to boot. My younger sister, Lisa Ann Poole, is always a generous source of companionship and support.

James Richard Keller had a bit of Richard Jordan Gatling in him. My late father was a mathematics professor at Marshall University, but the true passion of his life was two-fold: building things and dreaming about building things.

At the *Chicago Tribune,* I owe boundless and everlasting gratitude to Ann Marie Lipinski, not only as the editor, but also as my friend and mentor. Thanks are due as well to James Warren, Owen Youngman, Gerould Kern, Elizabeth Taylor, and Tim Bannon, along with my dear friends Emily Nunn and Heidi Stevens.

Longtime acquaintances are like stars in the night sky: You sometimes take their presence for granted, but then you look up and realize they have been lighting the way all along. Carolyn Focht, Virginia Rohan, Candy Justice, Jaye Bausser, Jennifer Crusie, Marlene Longenecker, Ruth Ann Hendrickson, and Kathleen Hill are deeply cherished, across many years and sometimes great distances, as are Marja Mills, Carla Mills, and Dave Mills. The friendship of John Phillips and Elaine

Phillips has been bright and sustaining. David Derbes, physics teacher extraordinaire, was kind enough to explain why a rifled barrel makes a bullet fly straighter and faster. And one gray day, deep in the catacombs of Princeton University's Firestone Library, when I despaired of finding that last obscure volume on the early history of firearms, it was the effervescence—and directional skills—of Carly Phillips that kept me going.

Susan Phillips, my best friend for more than two decades now, but for what feels like forever, was present at every step of this project, digging up arcane information about the nineteenth century and, best of all, sharing my admiration for Gatling and his inventions.

During the course of writing this book, I served as McGraw Professor of Writing at Princeton, where I relished conversations with Dr. Carol Rigolot. My residency there also enabled my friendship with Joyce Carol Oates, who has been an inspiration ever since.

Without the editing acumen of Wendy Wolf at Viking Penguin, this book would lack a large measure of whatever grace and coherence it now possesses. My agent, Stuart Krichevsky, has been a helpful guide through the wilds of the publishing world.

As mentioned in the notes below, this journey would have been far more daunting without the ready expertise and cheerful assistance of E. Frank Stephenson, Jr. I hope that Frank finds it worthy of the man whose name and achievements he has worked so hard to keep alive in American history: Richard Jordan Gatling.

NOTES

PAGE

ix *I have read a fiery gospel:* The first line of the third verse of Julia Ward Howe's famous poem is rarely quoted. Abraham Lincoln was a great fan of the song.

ix *They showed us:* Twain provided this account in a series of articles he wrote for a San Francisco newspaper, the *Alta California.* It was published March 3, 1868.

ix *"It's the Gatlings, men":* Theodore Roosevelt recounting the charge up San Juan Hill in *The Rough Riders*, p. 80.

INTRODUCTION

For most of his adult life, E. Frank Stephenson, Jr., has worked diligently on behalf of the memory of Richard Gatling. A professor at Chowan College in Murfreesboro, North Carolina, Frank grew up in the area; he spent his boyhood picking peanuts and hauling watermelon in the unbelievably intense Carolina sun. He befriended Gatling's grandson, John Waters Gatling, and Frank recently donated his large collection of Gatling-related materials to East Carolina University.

It was my great honor and good fortune to find Frank early in my researches, and he has been a friend and adviser all along the way. We met over hot dogs at Walt's Grill, a Murfreesboro hangout, and I've come back to him again and again with questions and observations. If there is an aspect of Gatling's early days with which Frank is unacquainted, it is surely not worth knowing. But I'd bet you a Walt's hot dog that no such aspect exists.

Frank's many books about the history of Murfreesboro and vicinity include a charming one called *Gatling: A Photographic Remembrance.* No student of Gatling's life and times should miss it.

The Gatling Gun, by Paul Wahl and Don Toppel, is, despite an exasperating lack of reference notes, an invaluable source of information about the gun's technical specifications and use in the Civil War and thereafter. Technical data also is available in the admirably lucid format of *The Gatling Gun Notebook*, by James B. Hughes. *The Social History of the Machine Gun*, by John Ellis, is one of the few books about firearms that considers their cultural function. It's a gem.

1 *The bravest men:* Quoted in Niall Ferguson, *The Cash Nexus: Money and Power in the Modern World, 1700–2000*, pp. 38–39. Rommel added, ". . . the gun's nothing without plenty of ammunition; and neither guns nor ammunition are of much use in mobile warfare unless there are vehicles with sufficient petrol to haul them around." In modern warfare, that is, equipment trumps valor.

2 *The discovery of quanta:* Robert Oppenheimer, "Robert Oppenheimer on Albert Einstein," in *The Company They Kept: Writers on Unforgettable Friendships*, ed. Robert B. Silvers and Barbara Epstein, p. 35.

2 *to hold a piece of death:* George Pelecanos, *The Sweet Forever*, p. 213.

4 *War, which every other nation:* Gary Wills, *Henry Adams and the Making of America*, p. 216.

5 *the sickle shape of the fallen:* Sebastian Barry, *A Long Long Way*, p. 185.

6 *shoveled into mass graves:* David Stevenson, *Cataclysm: The First World War as Political Tragedy*, p. 443.

6 *the Western European belligerents:* Stevenson, p. 443.

7 *It occurred to me:* Letter of June 15, 1877, from Richard Gatling to Lizzie Jarvis; the letter is part of the collection at the Lincoln Museum in Fort Wayne, Indiana.

8 *One couldn't pin:* John Ellis, *The Social History of the Machine Gun*, p. 18.

8 *unmilitary gimmick:* Ellis, p. 117.

8 *a crisscrossed lattice of death:* Quoted in Jasper Copping, "Mystery of Great War's Lost Army Uncovered," *London Sunday Telegraph*, July 22, 2007.

9 *In the early stages:* Quoted in Paul Avrich, *The Haymarket Tragedy*, p. 217.

9 *Do you know that the military:* Avrich, p. 202.

10 *"aristocratic lace wars'":* Paddy Griffith, *Battle Tactics of the Civil War*, p. 24.

10 *The sand of the desert:* Sir Henry Newbolt, "Vitai Lampada," *Oxford Book of War Poetry*, p. 243.

11 *that American poetry of vivid purpose:* William Dean Howells, *The Rise of Silas Lapham*, p. 70.

12 *life beyond self:* George Eliot, *Middlemarch*, p. 1.

12 *ushered in a regime:* Thomas Kessner, *Capital City: New York City and the Men Behind America's Rise to Economic Dominance, 1860–1900*, p. 241.

14 *It was during this period:* William Cronon, *Nature's Metropolis: Chicago and the Great West*, pp. xv–xvi.

14 *promoted the notion:* Charles Townshend, ed., "The Shape of Modern War," in *The Oxford History of Modern War*, p. 89.

15 *I regret to learn:* Letter of June 18, 1872, from Richard Gatling to L. W. Broadwell; the letter is in the Connecticut State Library collection.

CHAPTER ONE: **COLD BEAUTY**

Information on Governor O. P. Morton was gleaned from the Indiana Historical Bureau. The reports Morton received from the field of the killing and maiming of Indiana's young men came from the Indiana Commission on Public Records, which maintains a lengthy and comprehensive file of Morton's correspondence while holding public office. Background on the Civil War came from such books as James McPherson's elegant and justly praised *Battle Cry of Freedom* and the many fine Lincoln biographies, including *Lincoln*, by David Herbert Donald and Doris Kearns Goodwin's *Team of Rivals*. Other first-rate books on the Civil War are enumerated in the notes for chapter five.

17 *Nothing except a battle lost:* Quoted in Richard Holmes, *Wellington: The Iron Duke*, p. xvi.

17 *Take an even dozen:* James M. McPherson, *Battle Cry of Freedom: The Civil War Era*, p. 275.

17 *Yet the swift and sure:* Michael W. Kauffman, *American Brutus: John Wilkes Booth and the Lincoln Conspiracies*, p. 124.

17 *To His Excellency Governor Morton:* These and subsequent excerpts from casualty reports to Morton are contained in the O. P. Morton collection in the Indiana State Archives.

19 *had from thirty to forty per cent:* U. S. Grant, *Personal Memoirs of U. S. Grant*, p. 145.

19 *supremacy in firearms:* Angus Konstam, *The Pocket Book of Civil War Weapons*, pp. 40–41.

19 *Twenty-four hundred men in camp:* McPherson, p. 322.

20 *It is the opinion:* Robert V. Bruce, *Lincoln and the Tools of War*, p. 38.

20 *embarrassment in transacting:* Bruce, p. 39.

20 *superannuated, fretful, and slow:* Bruce, p. 39.

20 *Six months into the war:* Bruce, p. 57.

20 *In the first few months after Fort Sumter:* David A. Armstrong, *Bullets and Bureaucrats: The Machine Gun and the United States Army 1861–1916*, p. 10.

20 *A visitor to Washington, D.C.:* Jeffrey D. Wert, *The Sword of Lincoln: The Army of the Potomac*, p. 1.

21 *We are coming, Father Abraham:* James Sloan Gibbons, "Three Hundred Thousand More," in *Civil War Poetry: An Anthology*, edited by Paul Negri, p. 11.

22 *One enraged Democrat:* Jennifer L. Weber, *Copperheads: The Rise and Fall of Lincoln's Opponents in the North*, p. 81.

22 *That sort of indomitable drive:* Dee Brown, *The Year of the Century: 1876*, p. 79.

24 *Sir—Allow me:* Paul Wahl and Don Toppel, *The Gatling Gun*, p. 18.

25 *If you look up all our valleys:* Gibbons, p. 11.

25 *I have always thought:* Robert Oppenheimer, during congressional testimony on April 16, 1954, quoted in Kai Bird and Martin J. Sherwin, *American Prometheus: The Triumph and Tragedy of J. Robert Oppenheimer*, p. 520.

26 *the first great American invention:* Roger Ford, *The World's Great Machine Guns: From 1860 to the Present Day*, p. 19.

27 *some 80 percent of all casualties:* John Ellis, *The Social History of the Machine Gun*, p. 142.

27 *In 1861, during the opening:* Letter of June 15, 1877, from Richard Gatling to Lizzie Jarvis. The letter is in the collection of the Lincoln Museum in Fort Wayne, Indiana.

28 *Men of science:* William Hosley, *Colt: The Making of an American Legend*, p. 70.

28 *[Gatling guns] are so light:* Quoted in Wahl and Toppel, p. 19.

29 *Aaron Burr tried his hand:* Kirkpatrick Sale, *The Fire of His Genius: Robert Fulton and the American Dream*, p. 28.

29 *Vickers Brothers:* Clive Trebilcock, *The Vickers Brothers: Armaments and Enterprise, 1854–1914*, p. 26.

30 *a mountain howitzer:* Witold Rybczynski, *A Clearing in the Distance*, p. 139.

30 *probably saw the loan:* Jean Strouse, *Morgan: American Financier*, pp. 94–95.

31 *overnight cities:* McPherson, p. 17.

32 *When President Polk:* Bernard DeVoto, *The Year of Decision: 1846*, pp. 215–16.

32 *highly creditable to the genius:* DeVoto, p. 216.

32 *Pistols and false teeth:* Jacques Barzun, *From Dawn to Decadence, 1500 to the Present: 500 Years in Western Cultural Life*, pp. 553–55.

33 *Colt took apart:* William McNeill, *The Pursuit of Power*, pp. 233–34.

34 *In 1867, the International Exposition:* Kenneth Silverman, *Lightning Man: The Accursed Life of Samuel F. B. Morse*, pp. 420–21.

34 *Overwhelming all other exhibits, though:* Barbara Freese, *Coal: A Human History*, pp. 129–30.

34 *An athlete of steel:* Howard Mumford Jones, *The Age of Energy: Varieties of American Experience, 1865–1915*, p. 142.

35 *"Ice lemonade!":* Brown, pp. 129–32.

35 *"Whilst proud":* Brown, p. 130.

36 *great gallery of machines:* Henry Adams, *The Education of Henry Adams*, p. 380.

36 *had translated himself:* Adams, p. 381.

38 *improved existing models:* Alan Trachtenberg, *The Incorporation of America: Culture and Society in the Gilded Age*, p. 68.

38 *Dr. Morse's Indian Root Pills:* Robert B. Shaw, *History of Comstock Patent Medicine*, p. 10.

39 *Dr. Coult:* Hosley, p. 15.

39 *Electro-Magnetic Insulators:* Bruce, pp. 182–83.

39 *"I realize . . . that not nature alone":* Walt Whitman, "Democratic Vistas," quoted in Thomas Kessner, *Capital City: New York City and the Men Behind America's Rise to Economic Dominance, 1860–1900*, p. 89.

40 *the technological sublime:* David Nye, *American Technological Sublime.*

43 *bears the same relation:* Ellis, p. 16.

44 *From a marshy river bottomland:* Much of the information on Miles Greenwood and the early days of Cincinnati derives from *Centennial*

History of Cincinnati and Representative Citizens, by Charles Theodore Greve, first published in 1904, and Daniel Aaron's *Cincinnati: Queen City of the West, 1819–1838.*

44 *a special duty:* Siegfried Giedion, *Mechanization Takes Command*, p. 90.

45 *All that there is of good or bad:* Quoted in Debby Applegate, *The Most Famous Man in America: The Biography of Henry Ward Beecher*, p. 110.

46 *It had a square fire-box:* Greve, p. 659.

52 *"Mechanical construction is very simple":* Wahl and Toppel, p. 20.

52 *The bottom is out of the tub:* Stephen W. Sears, *Landscape Turned Red: The Battle of Antietam*, p. 28.

CHAPTER TWO: **A World of Mornings**

Much of the information about Gatling's early life comes from *Gatling: A Photographic Reminiscence*, by E. Frank Stephenson, Jr. Information about Gatling and the screw propeller is found in an article by Thomas C. Parramore entitled "The North Carolina Background of Richard Jordan Gatling." This essay, published in the *North Carolina Historical Review*, is distinguished not only by Parramore's research, but also by his passionate argument that his home state has unfairly neglected Gatling's memory because of old grudges that have festered since the Civil War.

Information on the lyceum movement is presented with thoroughness and historical perspective in Angela G. Ray's *The Lyceum and Public Culture in the Nineteenth-Century United States.* And the fascinating tale of the Lunar Club, philosophical precursor to so many of the learned societies in the United States, is found in Jenny Uglow's wonderful book *The Lunar Club.* Details about the War of 1812 come from A. J. Langguth's sprightly and readable *Union 1812: The Americans Who Fought the Second War of Independence.*

The history of the patent office jumps to life in the pages of *The Patent Office Pony: A History of the Early Patent Office*, by Kenneth Dobyns. General information about the U.S. Patent Office and the history of patent laws and famous patents comes from government publications: *The Story of the United States Patent Office*, published by the U.S. Department of Commerce in 1964 and 1971, and *The Story of the American Patent System 1790–1940*, published by the Commerce Department in 1940. To get a sense of the breadth and wondrousness of American ingenuity, two books—lavishly illustrated with photographs of inventors' models from the early Patent Office—are especially helpful: *The Art of Invention*, by William and Marlys Ray, and *American Enterprise: Nineteenth-Century Patent Models*, published by the Cooper-Hewitt Museum.

Patents were the great engine of nineteenth-century invention, but how are patents regarded today? Alas, many people believe Jefferson had it right all along: Ultimately, patents stifle innovation. Instead of protecting the rights of inventors, they end up enabling corporations to profit from legal monopolies, shutting newcomers out of the inventive process. The twenty-first-century world has begun to doubt the necessity of patents, to question the work of today's Patent Office, which is known as the U.S. Patent and Trademark Office. "The patent as stimulant to invention has long since given way to the patent as blunt instrument for establishing an innovation stranglehold," writes attorney Gary L. Reback in *Forbes*. "In fact, every patent issued comes at significant economic cost. . . . Too many patents are just as bad for society as too few." And in a world in which digitized information is flung around the globe in seconds, patents and copyrights seem like clumsy ankle weights attached to sleek runners. Have patents simply run their course? Do they belong to a fundamentally different era?

A 2006 series in the *Toledo Blade* argues otherwise. The articles highlight the research of two employees of the Cleveland Federal Reserve Bank. Analyzing why the state of Ohio has fallen behind other states in significant economic categories, the pair notes that in the past seventy-five years, the state has lost its edge in securing patents. "Patents per capita," according to the *Blade* article, "are the most important predictor of state wealth." In 1954, Ohio ranked sixth among all states for per-capita patent generation. In 1988, it had dropped to eleventh. By 2001, it came in at twentieth. "A wealth of patents drove Ohio's economy in the twentieth century," the *Blade* article states. "But in the past two decades, other states sprinted past."

To be sure, the increasing complexity of the world means that patents are no longer a matter of inspired amateurs beavering away in solitude and hope. As the innovation director of a medical center told the *Blade* reporters, "By and large, what we're doing is not going to be done in a garage." In the nineteenth century, many of the great discoveries did happen in barns and sheds, a result of the passion and drive of isolated individuals. For the space of one extraordinary century, these industrious entrepreneurs did their work, aspiring to procure patents. They came up with such things as telephones and vulcanized rubber and steamboats and pneumatic tires and dynamite and machine guns. Not a bad haul for mere amateurs.

53 *As it is:* Theodore Roosevelt's speech in Chicago, "Strenuous Life." Quoted in Sarah Watts, *Rough Rider in the White House: Theodore Roosevelt and the Politics of Desire*, p. 21.

54 *Between 1815 and 1839:* Paul Johnson, *The Birth of the Modern: World Society 1815–1830*, p. 218.

54 *the establishment:* Richard Slotkin, *The Fatal Environment: The Myth of the Frontier in the Age of Industrialization, 1800–1890*, p. 112.

55 *It's an age:* After writing this passage, I came across a marvelous description of the transformative impact of the telegraph in Kenneth Silverman's *Lightning Man: The Accursed Life of Samuel F. B. Morse:* "Morse had transmuted Thought, abstract human Thought, into metal strips and jars of acid," p. 241.

55 *Richard Gatling's father:* E. Frank Stephenson, Jr., *Gatling: A Photographic Remembrance*, p. 88.

58 *scraggly pine groves:* William Styron, *The Confessions of Nat Turner*, p. 237.

60 *wide-awake, every minute:* William Dean Howells, *The Rise of Silas Lapham*, p. 119.

61 *the noblest of Washington buildings:* Quoted in Kenneth W. Dobyns, *The Patent Office Pony: A History of the Early Patent Office*, p. 163.

62 *"The United States":* Thomas Kessner, *Capital City: New York City and the Men Behind America's Rise to Economic Dominance, 1860–1900*, p. 310.

62 *a country without a patent office:* Mark Twain, *A Connecticut Yankee in King Arthur's Court*, p. 49.

63 *"The Patent System":* Quoted in "The Story of the United States Patent Office," a 1972 government pamphlet published by the U.S. Department of Commerce and updated periodically, p. 1.

63 *The issue of patents:* Quoted in "The Story of the United States Patent Office," p. 1.

68 *a* London Telegraph *correspondent:* Ernest Furgurson, *Freedom Rising: Washington in the Civil War*, pp. 240–41.

68 *All who love new & strange sights:* Henry Whipple, *Bishop Whipple's Southern Diary 1843–1844*, p. 64.

69 *models of John Ericsson's screw propeller:* Robert Bruce, *Lincoln and the Tools of War*, pp. 61–62.

69 *Charles Dickens calls:* Bruce, p. 60.

69 *Thomas Edison is granted:* Jean Strouse, *Morgan: American Financier*, p. 181.

69 *A minor invention:* Jill Jones, *Empires of Light: Edison, Tesla, Westinghouse, and the Race to Electrify the World*, p. 52.

70 *that worldly cloister:* Walter Isaacson, *Einstein: His Life and Universe*, p. 78.

70 *We said, "What is it":* Quoted in B. Zorina Khan, *The Democratization of Invention: Patents and Copyrights in American Economic Development*, p. 21.

71 *All creation is a mine:* Tom Wheeler, *Mr. Lincoln's T-Mails: The Untold Story of How Abraham Lincoln Used the Telegraph to Win the Civil War*, p. 7.

71 *With every decade:* Johnson, p. 573.

72 *These were the days:* Bruce, p. 65.

72 *A collective fervor:* Siegfried Giedion, *Mechanization Takes Command*, p. 40.

72 *argue questions of theology:* Joseph Frazier Wall, *Andrew Carnegie*, p. 365.

74 *a shared esteem:* Kessner, p. 242.

74 *is not surpassed:* Quoted in Kessner, p. 73.

74 *light the way:* William Hosley, *Colt: The Making of an American Legend*, p. 138.

75 *a passion for art:* Hosley, p. 166.

75 *mitigate the unsettling:* Hosley, 166.

75 *instruments of the cosmopolitanism:* Hosley, p. 166.

76 *Your mission:* Frederick Merk, *Manifest Destiny and Mission in American History*, p. 262.

76 *It seems to me:* Angela G. Ray, *The Lyceum and Public Culture in the Nineteenth-Century United States*, p. 194.

77 *Front seats: a few old folks:* Ray, p. 35.

77 *the measured footstep:* Ray, p. 173.

78 *Lecturing is gymnastics:* Geoffrey C. Ward, Dayton Duncan, and Ken Burns, *Mark Twain*, p. 167.

80 *a consciousness that human society:* Strouse, p. 177.

80 *"Everybody knows":* Debby Applegate, *The Most Famous Man in America: The Biography of Henry Ward Beecher*, p. 268.

80 *believes that heaven intended:* Niall Ferguson, *Colossus: The Rise and Fall of the American Empire*, p. xxviii.

81 *the paradox:* Niall Ferguson, *Colossus: The Rise and Fall of the American Empire*, p. 54.

81 *empire in denial:* Ferguson, p. viii.

81 *not made up of aristocrats:* Jenny Uglow, *The Lunar Men: Five Friends Whose Curiosity Changed the World*, p. xiv.

82 *moonlighting inventor:* Jill Lepore, *A Is for American: Letters and Other Characters in the Newly United States*, p. 145.

83 *natural amateur:* Aaron Sachs, *The Humboldt Current: Nineteenth-Century Exploration and the Roots of American Environmentalism*, p. 144.

83 *science and art:* Uglow, p. xviii.

84 *Though the patent office's:* Bruce, p. 64.

84 *1,200 volunteers from Rhode Island:* Bruce, p. 63.

85 *The vast area of the second story:* Walt Whitman, *Walt Whitman's Civil War*, p. 87.

86 *unofficial ministering angel:* Stephen W. Sears, *Landscape Turned Red: The Battle of Antietam*, p. 306.

86 *"I could not help thinking":* Dobyns, p. 156.

88 *sensitivity to the unknown:* E. L. Doctorow, *The Waterworks*, p. 41.

CHAPTER THREE: **Land of the Second Chance**

89 *Any man of push:* Jones, p. 104.

89 *The universe is sensitive:* Speech given by Wanamaker on November 6, 1911; quoted in William Leach, *Land of Desire: Merchants, Power, and the Rise of a New American Culture*, p. 34.

89 *The future lies that way:* Henry David Thoreau, "Walking," in *The Writings of Henry David Thoreau*, pp. 217–18.

90 *one of the largest stretches:* Barbara Freese, *Coal: A Human History*, p. 104.

90 *the continent's seedbed:* Erik Reece, *Lost Mountain*, p. 35.

90 *Always Americans:* Bruce Catton and William Catton, *Two Roads to Sumter*, p. 5.

91 *All your campfires:* Quoted in Tom Chaffin, *Pathfinder: John Charles Frémont and the Course of American Empire*, p. xxiv.

91 *frontier wilderness:* Catton and Catton, pp. 33–35.

91 *silence and the unknown:* Chaffin, p. 136.

92 *goods, wares and merchandize:* Certificate from 1840, from collection at East Carolina Manuscript Collection, Box 25, File B.

92 *Roaring Forties:* Bernard DeVoto, *The Year of Decision: 1848*, p. 8.

93 *One hapless man:* Paul Johnson, *The Birth of the Modern: World Society 1815–1830*, p. 171.

93 *To the New England mind:* Henry Adams, *The Education of Henry Adams*, p. 47.

93 *In 1846, when an actor:* Frederick Lewis Allen, "The Big Change," in *An American Retrospective*, ed. Ann Marie Cunningham, p. 35.

93 *from Cincinnati to Toledo:* Debby Applegate, *The Most Famous Man in America: The Biography of Henry Ward Beecher*, pp. 107–8.

94 *In benighted regions:* Harriet Beecher Stowe, *Uncle Tom's Cabin,* p. 104.

94 *one of the magic keys:* Johnson, p. 177.

94 *Of all inventions:* Garry Wills, *Henry Adams and the Making of America*, p. 128.

95 *Every piece of stone:* Johnson, p. 178.

95 *But our hearts are:* Herman Melville, *White Jacket; or, The World in a Man-of-War*, pp. 152–53; 160.

96 *Yes, more, more, more!:* Frederick Merk, *Manifest Destiny and Mission in American History*, p. 46.

96–97 *To live in such a splendid country:* Merk, p. 55.

97 *The patriotic impulses:* Merk, p. 46.

97 *"Go to the West":* Merk, p. 29.

99 *idealistic, self-denying:* Merk, p. 261.

100 *I became fully sensible:* Siegfried Giedion, *Mechanization Takes Command*, p. 143.

100 *The earth is a machine:* Ralph Waldo Emerson, "Farming," in *The Essential Writings of Ralph Waldo Emerson*, p. 676.

102 *good schools & churches:* Henry Whipple, *Bishop Whipple's Southern Diary, 1843–44,* p. 134.

102 *I can call back:* Geoffrey C. Ward, Dayton Duncan, and Ken Burns, *Mark Twain*, pp. 216–17.

102 *That's the sadness:* Larry McMurtry, *The Wandering Hill,* p. 387.

103 *so dense as to render it necessary:* Landon Y. Jones, *William Clark and the Shaping of the West*, p. 275.

104 *the meeting point:* Frederick Jackson Turner, "The Significance of the Frontier in American History," in *The American Intellectual Tradition,* edited by

David A. Hollinger and Charles Capper, p. 85.

104 *new men:* Adams, pp. 462–63.

105 *elastic vigor of youth:* Quoted in Michael Kammen, *Mystic Chords of Memory: The Transformation of Tradition in American Culture*, p. 57.

105 *This perennial rebirth:* Turner, p. 85.

105 *New communities:* Kammen, p. 51.

106 *Like thousands of other:* Ward, Duncan, and Burns, p. 3.

106 *held under their hammers:* Adams, p. 309.

108 *"Great God, Grant":* Jean Edward Smith, *Grant*, p. 91.

108 *Humiliated, bankrupt:* Edmund Wilson, *Patriotic Gore: Studies in the Literature of the American Civil War*, p. 144.

109 *a dirty place:* Allen, p. 37.

111 *glorious rival:* William Cronon, *Nature's Metropolis: Chicago and the Great West*, p. 297.

111 *The new gateway:* Cronon, p. 307.

112 *TO FARMERS:* The poster is contained in the E. Frank Stephenson, Jr. collection at East Carolina Manuscript Collection, J. Y. Joyner Library, East Carolina University.

CHAPTER FOUR: "DRUNKARDS, DANDIES & LOAFERS"

Louis C. Hunter's *Steamboats on the Western Rivers: An Economic and Technological History*, with its impeccably thorough statistics, is the indisputably essential source for that brief, colorful, and perilous period in American history when steamboats ruled the waters. Also helpful was a more obscure but equally fascinating book, *A-Rafting on the Missisip'*, by Charles Edward Russell, a delightful account of the river world of the mid-nineteenth century. When it comes to smallpox, *The Demon in the Freezer*, by Richard Preston, is a highly readable (and terrifying) account of the disease. A detailed account of the patent medicine business in nineteenth-century America can be found in Robert B. Shaw's *History of the Comstock Patent Medicine Business and Dr. Morse's Indian Root Pills.*

113 *The river persuades:* Russell Celyn Jones, *Ten Seconds from the Sun*, p. 24.

114 *You are besieged:* Henry Whipple, *Bishop Whipple's Southern Diary 1843–1844*, p. 122.

114 *Instead of a quiet and a reasonable number:* Whipple, p. 138.

114 *rushing down the Mississippi:* Louis C. Hunter, *Steamboats on the Western Rivers: An Economic and Technological History*, p. 29.

116 *I did not go much about town:* Whipple, p. 150.

117 *swept up that river:* Larry McMurtry, *By Sorrow's River*, pp. 139–40.

117 *the great sickness:* McMurtry, p. 149.

119 *the entire nation:* David Howard Bain, *Empire Express: Building the First Transcontinental Railroad*, p. 664.

119 *Steamboats took fire:* Charles Edward Russell, *A-Rafting on the Mississip'*, p. 32.

120 *Machinery, vast fragments:* Hunter, p. 288.

120 *I can never forget:* Russell, pp. 30–31.

120 *For forty-eight hours:* Ward, Duncan, and Burns, p. 20.

121 *No man will ever forget:* Quoted in Howard Mumford Jones, *The Age of Energy: Varieties of American Experience, 1865–1915*, p. 67.

121 *What a rush:* Whipple, pp. 138–39.

122 *floating palaces:* Kirkpatrick Sale, *The Fire of His Genius: Robert Fulton and the American Dream*, p. 183.

122 *Orange-women and news-boys:* Sale, p.183.

122 *The river had a new story:* Geoffrey C. Ward, Dayton Duncan, and Ken Burns, *Mark Twain*, p. 18.

122 *a wondrous interruption:* Ron Powers, *Mark Twain: A Life*, p. 30.

122 *It was a great, absorbing, dominating:* Russell, pp. 239–40.

123 *"You have undertaken":* "Introductory Lecture at the Opening of the Thirtieth Session of the Medical College of Ohio," November 5, 1849. Cincinnati: Morgan and Overend Printers, 1849. From the Ohio Historical Society.

125 *a small butterscotch-colored notebook:* Gatling's notebook is part of the E. Frank Stephenson collection at East Carolina University.

126 *self-instructed and self-certified:* Todd Timmons, *Science and Technology in Nineteenth-Century America*, p. 141.

126 *An American physician:* Timmons, p. 141.

127 *from the Maritime provinces:* Robert B. Shaw, *History of the Comstock Patent Medicine Business and Dr. Morse's Indian Root Pills*, p. 40.

127 *Department stores did not appear:* William Leach, *Land of Desire: Merchants, Power, and the Rise of a New American Culture*, p. 20.

128 *healthy blooming family:* Shaw, p. 10.

128 *was the first man:* Shaw, p. 11.

128 *Dr. Larzetti's Juno Cordial:* Shaw, p. 29.

128 *They have an especial action:* Shaw, p. 29.

129 *"Experience, down to the present hour":* The address by Dr. Drake is contained in unpublished records maintained by the Ohio Historical Society.

130 *It seems that Senator John Scott Harrison:* Pauline Chen, *Final Exam: A Surgeon's Reflections on Mortality*, p. 24.

130 *biliousness; dyspepsia:* Shaw, p. 32.

132 *Two Kentuckians:* Bruce Catton and William Catton, *Two Roads to Sumter*, pp. 2–3.

132 *lived in the family:* U. S. Grant, *Personal Memoirs of U. S. Grant*, pp. 4–5.

132 *"It seems," a dying Ulysses S. Grant:* Quoted in Edmund Wilson, *Patriotic Gore: Studies in the Literature of the American Civil War*, p. 138.

133 *The whole city was given over:* Debby Applegate, *The Most Famous Man in America: The Biography of Henry Ward Beecher*, pp. 164–65.

133 *begins to look like a town:* Paul Johnson, *The Birth of the Modern: World Society 1815–1830*, p. 217.

139 *"I wish I had invented":* Quoted in Larry Kahaner, *AK-47: The Weapon That Changed the Face of War*, p. vi.

CHAPTER FIVE: **The Spaces Between the Bullets**

Dante and Shakespeare split up the world, T. S. Eliot once opined in an essay in *The Sacred Wood.* That is how I feel about Bruce Catton and Shelby Foote, when it comes to the Civil War. There are many, many books about the Civil War, but these two writers truly seem to have inhabited it, body and soul. My understanding of the war—not just its numbers and historical implications, but the feel of it, the sweep of its narrative—has been deepened, enriched, and illuminated at every turn by their magnificent books. This chapter also owes a heavy debt to Robert Bruce's remarkable *Lincoln and the Tools of War,* which restores a much-needed balance to the portrait of a president whom most people think they know so well. David Armstrong's *Bullets and Bureaucrats* is an irrefutable indictment of American military leadership's shortsightedness when it came to machine guns.

141 *Confederates who beheld it:* Bruce Catton, *Reflections on the Civil War*, p. 128.

142 *so awful ugly:* Walt Whitman, *Walt Whitman's Civil War*, p. 174.

142 *a fountain of first-class:* Whitman, p. 174.

142 *He hunted down:* Shelby Foote, *The Civil War: A Narrative: Fort Sumter to Perryville*, p. 142.

143 *"Intellect shone through":* Joseph Frazier Wall, *Andrew Carnegie*, p. 168.

144 *in a society of hunters:* William Lee Miller, *Lincoln's Virtues: An Ethical Biography*, p. 43.

145 *ushered immediately:* Robert V. Bruce, *Lincoln and the Tools of War*, p. 142.

145 *During the Civil War:* Bruce, p. 225.

146 *an army in six feet square:* Paul Wahl and Don Toppel, *The Gatling Gun*, p. 10.

149 *a machine in which:* John Ellis, *The Social History of the Machine Gun*, p. 11.

150 *that discharges so often:* Ellis, p. 13.

151 *Vandenburgh volley gun:* These descriptions come from Angus Konstam, *The Pocket Book of Civil War Weapons*, pp. 202–09.

151 *the devil's paintbrush . . . potato digger:* Roger Ford, *The World's Great Machine Guns From 1860 to the Present Day*, p. 41.

152 *inefficient, and unsafe to operate:* David A. Armstrong, *Bullets and Bureaucrats: The Machine Gun and the United States Army, 1861–1916*, p. 19.

153 *"I saw one of them work":* Wahl and Toppel, p. 10.

153 *the machine-gun . . . enables[s]:* Martin Van Creveld, "Technology and War I," in *The Oxford History of Modern War*, p. 216.

154 *Presently he proceeded:* Stephen Crane, *The Red Badge of Courage*, p. 48.

155 *I have seen an inferior arm:* Wahl and Toppel, p. 22.

156 *almost two centuries:* Paddy Griffith, *Battle Tactics of the Civil War*, p. 76.

157 *"arm was all battered":* Griffith, p. 84.

159 *pop-popping:* Foote, p. 199.

159 *raised a shout:* Richard Holmes, *Wellington: The Iron Duke*, p. 116.

159 *I cannot give you particulars:* Bruce Catton, *Terrible Swift Sword*, p. 434.

160 *The rattle of machine-guns:* Erich Maria Remarque, *All Quiet on the Western Front*, p. 216.

160 *very white hair:* Bruce, p. 40.

161 *a great evil:* Bruce, p. 69.

161 *The sense that victory:* Van Creveld, p. 206.

161 *Dead battles, like dead generals:* Quoted in Anthony Smith, *Machine Gun: The Story of the Men and the Weapon That Changed the Face of War*, p. 171.

162 *"The army," charged military historian:* Armstrong, p. ix.

164 *The arm in question:* Wahl and Toppel, p. 22.

166 *"I saw General Butler":* Wahl and Toppel, p. 20.

166 *City is in intense excitement:* Jennifer L. Weber, *Copperheads: The Rise and Fall of Lincoln's Opponents in the North*, p. 108.

167 *tearing up rails:* Weber, p. 108.

168 *"Give them grape":* Wahl and Toppel, p. 24.

169 *"Glory!":* Bruce Catton, *Never Call Retreat*, p. 447.

170 *Lincoln's route led him:* Descriptions of Lincoln's walk through Richmond on April 4, 1865, and of Lee's surrender at Appomattox are gleaned from many sources, including Bruce Catton, *Never Call Retreat*, pp. 444–469, and U. S. Grant, *Personal Memoirs.*

170 *If there be any:* Catton, *Never Call Retreat*, p. 448.

CHAPTER SIX: "A LITTLE GATLING MUSIC"

An impressive portrait of Hartford emerges in William Hosely's *Colt: The Making of an American Legend*, a large-scale, lavishly illustrated book about the city, the Colt factory, and the way each influenced the other. The information about Buffalo Bill derives from the sources noted below, but to get you in the mood to write about that inimitable showman, you can't do better than Larry McMurtry's *The Colonel and Little Missie: Buffalo Bill, Annie Oakley, and the Beginnings of Superstardom in America.* Whether he's writing novels or history, McMurtry's love for the West—a love that doesn't blind him to its hardships or its negatives—shines forth.

In this chapter and the next, I also relied heavily on that towering work of scholarship, William McNeill's *The Pursuit of Power.* Professor McNeill not only seems to know everything worth knowing about armaments history, he tells what he knows in crisp, captivating language.

For a deeper understanding of Theodore Roosevelt and his singular psychology, I utilized the many Roosevelt biographies, as well as Sarah Watts's lively and learned *Rough Rider in the White House: Theodore Roosevelt and the Politics of Desire.* I don't agree with all of her conclusions, but this well-researched, sharply argued account provides a new (and rather disturbing) way to look at Roosevelt and his hypermasculine exploits in a nation unsure of its identity as a new century loomed.

173 *I remember the ceaseless:* Richard Harding Davis, *Captain Macklin*, p. 272.

173 *97 Indians:* Robert W. Rydell and Rob Kroes, *Buffalo Bill in Bologna: The Americanization of the World*, 1869–1922, p. 106.

175 *a Gatling gun on paper:* Editorial, *New York Times*, July 3, 1894. Quoted in Almont Lindsey, *The Pullman Strike: The Story of a Unique Experiment and of a Great Labor Upheaval*, p. 162.

175 *treat them to a little Gatling music:* Quoted in John Ellis, *The Social History of the Machine Gun*, p. 82.

176 *every man's position:* Robert W. Rydell and Bob Kroes, *Buffalo Bill in Bologna: The Americanization of the World, 1869–1922*, pp. 114–15.

177 *"a wonderful little girl":* Rydell and Kroes, p. 107.

177 *"All present were constrained":* Rydell and Kroes, p. 108.

178 *It is often said:* Rydell and Kroes, p. 113.

178 *The bullet:* Rydell and Kroes, p. 114.

178 *"Cody," the wily entrepreneur:* Robert A. Carter, *Buffalo Bill Cody: The Man Behind the Legend*, p. 310.

179 *After thorough and exhausting trials:* Letter of March 16, 1874, to the Secretary of War Recommending an Appropriation for Gatling Guns, from the Acting Chief of Ordnance, pp. 1–2; Connecticut State Library; Colt Collection.

182 *Colonel James Wolfe Ripley:* Robert V. Bruce, *Lincoln and the Tools of War*, p. 298

182 *lovely place:* Witold Rybczynski, *A Clearing in the Distance: Frederick Law Olmsted and America in the 19th Century*, p. 32.

182 *Of all the beautiful towns:* Geoffrey C. Ward, Dayton Duncan, and Ken Burns, *Mark Twain*, p. 88.

183 *a muddy, smoky:* Ward, Duncan, and Burns, p. 89.

183 *Between 1850 and 1960:* William Hosley, *Colt: The Making of an American Legend*, p. 55.

183 *a Silicon Valley of its day:* Anthony Smith, *Machine Gun: The Story of the Men and the Weapon that Changed the Face of War*, p. 50.

183 *The world, after 1865:* Henry Adams, *The Education of Henry Adams*, p. 247.

184 *no official lender:* Jean Strouse, *Morgan: American Financier*, p. 6.

184 *America's antiquated banking system:* Strouse, p. 6.

184 *the broadest, straightest streets:* Mark Twain, "Mark Twain on His Travels," *San Francisco Alta Vista,* March 3, 1868.

185 *a great range of tall brick buildings:* Twain, "Mark Twain on his Travels."

187 *Employed under conditions:* David A. Armstrong, *Bullets and Bureaucrats: The Machine Gun and the United States Army, 1861–1916*, p. 80.

188 *"clumsy" and "frightening":* Evan S. Connell, *Son of the Morning Star: Custer and the Little Bighorn*, p. 57.

188 *These weapons were invented:* Connell, p. 257.

188 *Mechanical construction:* Paul Wahl and Don Toppel, *The Gatling Gun*, p. 20.

188 *has stood the limited test:* Wahl and Toppel, p. 20.

188–189 *All parts of the gun work well:* Harvey Brandt, "Gatling Gun," *Gun Digest*, p. 22, citing C. B. Norton's *American Breech-Loading Small Arms*, p. 242.

189 *Certainly they acknowledged:* Ellis, pp. 49–50.

189 *I returned to this city:* Letter of October 14, 1868, from Richard Gatling to General John Love. Indiana Historical Society collection.

190 *international arms manufacturers:* William Manchester, *The Arms of Krupp: 1587–1968*, p. 223.

190 *Great states constitute:* Charles H. Foster, "The Modern Vulcan," *Potter's American Monthly* XII, no. 89 (May 1879): p. 333.

190 *When Krupp sold guns:* Foster, p. 184.

190 *being heavily dependent:* Neil McKendrik, introduction to *The Vickers Brothers: Armaments and Enterprise 1854–1914*, by Clive Trebilcock, p. xxxii.

191 *They lived in a period:* Clive Trebilcock, *The Vickers Brothers: Armaments and Enterprise, 1854–1914*, p. 152.

191 *Warriors didn't fight:* Manchester, p. 299.

191 *And Fritz Krupp:* Manchester, p. 236.

192 *Official [Russian] policy:* William McNeill, *The Pursuit of Power*, p. 259.

192 *Cheap machine-made goods:* McNeill, pp. 260–61.

192 *In 1898, Fritz Krupp gave:* Manchester, 257.

192 *six feet high [and] trimmed:* Hosley, p. 30.

193 *A Notice:* Quoted in Wahl and Toppel, p. 100.

194 *I regret to say:* Letter of February 8, 1895, from Richard Gatling to Rebecca Peebles; E. Frank Stephenson, Jr. Collection; East Carolina University.

195 *The savage does not fight:* Wahl and Toppel, p. 69.

195 *We are not surprised:* Wahl and Toppel, p. 69.

195 *When all was over:* Quoted in Ellis, p. 84.

196 *Round whisked the Gatlings:* Quoted in Ellis, p. 84.

196 *With machine guns in their armoury:* Ellis, p. 79.

196 *Machine guns were lengthily regarded:* Smith, p. 114.

196–197 *regarded the machine gun as a weapon:* Quoted in Ellis, p. 103.

197 *were not in favor of slavery:* Letter of June 12, 1957, from John Waters Gatling to William B. Edwards; E. Frank Stephenson, Jr. Collection, East Carolina University.

198 *one of the most fashionable events:* E. Frank Stephenson, Jr., *Gatling: A Photographic Remembrance*, p. 62.

198 *Banks of flowers:* Quoted in Stephenson, p. 62.

200 *It is needless to say:* Letter of August 9, 1884, from Richard Gatling to Rebecca Peebles; E. Frank Stephenson, Jr. Collection, East Carolina University.

201 *If you knew how sadly:* Letter of November 22, 1897, from Frank Dixon to Jemima Gatling; E. Frank Stephenson, Jr. Collection, East Carolina University.

201 *that heaven intended it:* Niall Ferguson, *Colossus: The Rise and Fall of the American Empire*, p. xxviii.

202 *You could not impose:* Letter of June 21, 1957, from John W. Gatling to William B. Edwards; E. Frank Stephenson, Jr. Collection; East Carolina University.

202 *accomplish on the farm:* "New Channel for Dr. Gatling's Genius," *St. Louis Republic*, August 5, 1901.

203 *You are lovely:* Undated telegram from Ida Gatling to Jemima Gatling; E. Frank Stephenson, Jr. Collection, East Carolina University.

203 *My Dear Son Richard:* Letter of November 21, 1902, from Richard Gatling to his son Richard; E. Frank Stephenson, Jr. Collection, East Carolina University.

204 *As he turned on his heel:* Theodore Roosevelt, *The Rough Riders*, pp. 73–74.

205 *were trained by life-long habit:* Roosevelt, p. 62.

205 *"It's the Gatlings":* Roosevelt, p. 80.

205 *We saw much of Parker:* Roosevelt, p. 80.

205 *Indeed, the dash:* Roosevelt, p. 87.

206 *we usually got Parker:* Roosevelt, p. 98.

206 *the blood-bought victory:* Roosevelt, p. 98.

CHAPTER SEVEN: "THE WORLD'S GREAT STORM"

So many fine books abound on World War I that selecting only a few to acknowledge is daunting. Even though it is now more than four decades old, Barbara Tuchman's *The Guns of August* is an essential narrative about the political buildup to the war. Her selection of details and her ability to weave individual personalities into the larger pattern of events are unsurpassed. Among more recent works, I am especially indebted to David Stevenson's *Cataclysm: The First World War as Political Tragedy* and G. J. Meyer's *A World Undone: The Story of the Great War 1914 to 1918.* The latter is a wonderfully accessible account of a highly complex conflict.

Fiction and poetry supply the unforgettable images of this most visceral of wars, from classics such as *All Quiet on the Western Front,* by Erich Maria Remarque, to the poems of Wilfred Owen and Siegfried Sassoon. A more recent novel, Sebastian Barry's *A Long Long Way,* is not to be missed.

207 *Bombardment, barrage:* Erich Maria Remarque, *All Quiet on the Western Front,* p. 132.

207 *He grew used to the sight:* Sebastian Faulks, *Birdsong,* p. 162.

209 *Dr. Gatling Dead:* Several Gatling obituaries are found in E. Frank Stephenson, Jr.'s *Gatling: A Photographic Remembrance,* pp. 74–82.

210 *For many years:* The text of the Gatling eulogy is contained in Stephenson, p. 83.

212 *If we are to be crushed:* Tuchman, *The Guns of August,* p. 124.

213 *the world's great storm:* Joseph Conrad, dedication to *The Rescue,* dedication page.

213 *scientists, engineers and mechanics:* John Bourne, "Total War I: The Great War," in *The Oxford History of Modern War,* ed. Charles Townshend, p. 131.

213 *When we started to fire:* G. J. Meyer, *A World Undone: The Story of the Great War, 1914 to 1918,* p. 376.

213 *the spell of élan:* Tuchman, p. 77

214 *a much-overrated weapon:* Meyer, p. 379.

214 *Our machine guns did excellent work:* Meyer, p. 192.

214–215 *"As war was, so it has remained":* Tuchman, p. 80.

215 *the principles of strategy:* Tuchman, p. 80.

215 *It kept getting between my legs:* Theodore Roosevelt, *The Rough Riders*, p. 57.

216 *negated all the old human virtues:* John Ellis, *The Social History of the Machine Gun*, pp. 16–17.

216 *sneaky and un-Nelsonian:* William McNeill, *The Pursuit of Power*, p. 298.

216 *the old notion:* McNeill, p. 298.

217 *A rifle bullet:* Robert Graves, quoted in Anthony Smith, *Machine Gun: The Story of the Men and the Weapon that Changed the Face of War*, p. 187.

217 *A weapon that could be used:* McNeill, p. 172.

218 *Reservists went:* Tuchman, pp. 94–95.

218 *the blow that hurled:* Jacques Barzun, *From Dawn to Decadence: 1500 to the Present: 500 Years of Western Cultural Life,* p. 683.

220 *"glades and lawns":* Roger Ford, *The World's Great Machine Guns*, p. 31.

221 *chronic inventor:* Smith, p. 81.

221 *exquisite:* Smith, p. 85.

222 *"Hang your chemistry":* Smith, p. 82.

223 *It occurred to me:* Richard Gatling's letter of June 15, 1877, to Lizzie Jarvis.

223 *industrialize war:* Niall Ferguson, *The Cash Nexus: Money and Power in the Modern World, 1700–2000*, pp. 49–50.

223 *Technology has lowered:* Ferguson, p. 38.

224 *The destructive effect:* John H. Parker, *Tactical Organization and Uses of Machine Guns in the Field*, p. 157.

224 *It is confidently believed:* Gatling's 1863 sales brochure, reproduced in Stephenson, p. 44.

CHAPTER EIGHT: WARRIORS AND SAGES

My visits to Murfreesboro, North Carolina, were fascinating—and hot. Standing in the Gatling family cemetery, in the middle of a cotton field, at noon in mid-July is every bit as oozingly uncomfortable and enervating as it sounds. Yet when I reflected that I was merely standing there taking notes, while generations of North Carolinians have toiled for many hours each day in that same heat, I decided to stop complaining. Frank Stephenson was a charming and knowledgeable guide to every corner of Murfreesboro, indeed to the whole of Hertford County, as well as to all things Gatling. He took me to the local historical museum, which features a Gatling gun that Frank had personally procured; the Jeffcoat Museum, with its amazing multifloored display of wringer washers and phonographs and thousands of examples of ingenious Americana; and to that cemetery, a place that encourages contemplation about things past and present.

I've been to the Gatling family plot at Crown Hill Cemetery several times. Each visit is an adventure. John Dillinger is buried at the cemetery; so is President Benjamin Harrison. You never know just who you'll run into in a graveyard. Crown Hill is a lovely place, especially the part with the Civil War graves; they rise along a small hill shaded by ancient trees. Those trees look as if they're meditating on what lies beneath and above. No matter what your state of mind, you're bound to leave Crown Hill feeling calmer.

225 *A multitude of simultaneous deaths:* Quoted in Marianna Torgovnick, *The War Complex: World War II in Our Time*, p. xi.

225 *A few Sokolov Maxim:* William T. Vollmann, *Europe Central*, p. 268.

225 *As a practical military machine gun:* The card is reproduced in James B. Hughes, *The Gatling Gun Notebook*, p. 85.

226 *Do not forget:* Quoted in Paul Wahl and Don Toppel, *The Gatling Gun*, p. 75.

227 *in the process of rapid development:* Richard Slotkin, *The Fatal Environment: The Myth of the Frontier in the Age of Industrialization, 1800–1890*, p. 32.

228 *one of the foremost:* Charles H. Foster, "The Modern Vulcan." *Potter's American Monthly,* XII, 89, May 1879, p. 321.

228 *As every advance in knowledge:* Leo Braudy, *The Frenzy of Renown*, p. 553.

229 *vehicles of cultural memory:* Braudy, p. 15.

229 *where personal psychology:* Braudy, p. 16.

229 *the old Gatling homestead:* Foster, p. 329.

229 *Not a leaf:* Foster, p. 330.

231 *We have seen him pass:* Foster, p. 333.

232 *were no more self-confident:* Niall Ferguson, *Colossus: The Rise and Fall of the American Empire*, p. 33.

232 *"nascent empire" and "infant empire":* Ferguson, p. 34.

233 *The Industrial Revolution in Europe:* Geoffrey Parker, *The Military Revolution: Military Innovation and the Rise of the West*, p. 4.

233 *Once the Industrial Revolution:* Parker, pp. 174–75.

233 *The improvements in the muzzle-loading guns:* Paul Kennedy, *The Rise and Fall of the Great Powers*, p. 150.

234 *Literate societies with metal tools:* Jared Diamond, *Guns, Germs, and Steel: The Fates of Human Societies*, p. 13.

234 *The whole modern world:* Diamond, p. 25.

234 *Technology, in the form of:* Diamond, p. 241.

234 *several constellations of:* Diamond, p. 31.

234 *at most, a reluctant imperialist:* John Steele Gordon, *An Empire of Wealth: The Epic History of American Economic Power*, p. xiii.

235 *"My problem," he wrote:* Ron Chernow, *Titan: The Life of John D. Rockefeller, Sr.*, p. 509.

235 *Rockefeller's life:* Chernow, p. xv.

236 *The development of a true* mass *media:* Slotkin, p. 32.

236 *the application of celebrity:* Randall Stross, *The Wizard of Menlo Park: How Thomas Alva Edison Invented the Modern World*, p. 1.

237 *near-volcanic change:* Alan Trachtenberg, *The Incorporation of America: Culture and Society in the Gilded Age*, p. xiv.

237 *The axis on which:* Trachtenberg, p. ix.

238 *heady, adolescent, picturesque:* Howard Mumford Jones, *The Age of Energy Varieties of American Experience, 1865–1915*, p. 434.

238 *the march of progress:* Paul Johnson, *Birth of the Modern: World Society 1815–1830*, p. 813.

238 *a positive force:* Johnson, p. 814.

238 *"History," William James reminded:* William James, *Memories and Studies*, p. 269.

239 *"What we want to do":* Linton Brooks, quoted in multiple stories, including the *Oakland Tribune*, February 3, 2006.

BIBLIOGRAPHY

Institutions

Cincinnati History Museum, Cincinnati, Ohio
Connecticut State Library, Hartford, Connecticut
Firestone Library, Princeton University, Princeton, New Jersey
Indiana Commission on Public Records, Indianapolis, Indiana
Indiana State Historical Society, Indianapolis, Indiana
J. Y. Joyner Library, East Carolina University, Greenville, North Carolina
Lincoln Museum, Fort Wayne, Indiana
National Archives, Washington, D.C.
Newberry Library, Chicago, Illinois
New York Public Library, New York, New York
Ohio Historical Society, Columbus, Ohio

Books and Articles

Aaron, Daniel. *Cincinnati: Queen City of the West, 1819–1838.* Columbus: Ohio State University Press, 1992.

Adams, Henry. *The Education of Henry Adams.* New York: Modern Library, 1931.

Allen, Frederick Lewis. "The Big Change: The Coming and Disciplining of Industrialism 1850–1950." In *An American Retrospective*, edited by Ann Marie Cunningham, 34–41. New York: Harper's Magazine Foundation, 1984.

Amis, Martin. *House of Meetings.* New York: Alfred A. Knopf, 2007.

Andersen, Kurt. *Heyday.* New York: Random House, 2007.

Applegate, Debby. *The Most Famous Man in America: The Biography of Henry Ward Beecher.* New York: Three Leaves Press, 2006.

Armstrong, David A. *Bullets and Bureaucrats: The Machine Gun and the United States Army, 1861–1916.* Westport, CT: Greenwood Press, 1982.

Atkins, Peter. *Galileo's Finger: The Ten Great Ideas of Science.* London: Oxford University Press, 2003.

Avirch, Paul. *The Haymarket Tragedy.* Princeton, NJ: Princeton University Press, 1984.

Ayers, Edward L. *In the Presence of Mine Enemies: War in the Heart of America, 1859–1863.* New York: W. W. Norton, 2003.

Bain, David Howard. *Empire Express: Building the First Transcontinental Railroad.* New York: Viking, 1999.

Bakeless, John, ed. *The Journals of Lewis and Clark.* New York: Signet, 2002.

Barry, Sebastian. *A Long Long Way.* New York: Penguin, 2006.

Barzun, Jacques. *From Dawn to Decadence: 1500 to the Present: 500 Years in Western Cultural Life.* New York: HarperCollins, 2000.

Beaumont, Richard. *Purdey's: The Guns and the Family.* North Pomfret, VT: David & Charles, 1984.

Bellamy, Edward. *Looking Backward.* New York: Penguin, 1982.

Bellesiles, Michael A. *Arming America: The Origins of a National Gun Culture.* New York: Vintage, 2000.

Benfrey, Christopher. "American Jeremiad." *New York Review of Books*, September 22, 2005, 65–67.

Biel, Steven. *Down with the Old Canoe: A Cultural History of the Titanic Disaster.* New York: W. W. Norton, 1996.

Bird, Kai, and Martin J. Sherwin. *American Prometheus: The Triumph and Tragedy of J. Robert Oppenheimer.* New York: Vintage, 2006.

Bodanis, David. *Electric Universe: The Shocking True Story of Electricity.* New York: Crown, 2005.

Boomhower, Ray E. *The Sword & the Pen: A Life of Lew Wallace.* Indianapolis: Indiana Historical Society Press, 2005.

Bourne, John. "Total War I: The Great War." In *The Oxford History of Modern War*, edited by Charles Townshend, 117–37. New York: Oxford University Press, 2005.

Brandt, Harvey. "Gatling Gun." In *Gun Digest*, edited by John T. Amber, 20–29. Chicago: Gun Digest Co., 1957.

Braudy, Leo. *The Frenzy of Renown.* New York: Oxford University Press, 1986.

Brittain, Vera. *Testament of Youth.* New York: Wideview Books, 1980.

Brown, Dee. *The Year of the Century: 1876.* New York: Charles Scribner's Sons, 1966.

Brown, G. I. *The Big Bang: A History of Explosives.* Phoneix Mill, England: Sutton Publishing, 1998.

Browne, Janet. *Charles Darwin: The Power of Place.* Princeton, NJ: Princeton University Press, 2002.

Bruce, Robert V. *Lincoln and the Tools of War.* Indianapolis: Bobbs-Merrill, 1956.

Buchanan, Brenda J. *Gunpowder, Explosives and the State: A Technological History*. Burlington, VT: Ashgate Publishing, 2006.

Burton, Orville Vernon. *The Age of Lincoln*. New York: Hill and Wang, 2007.

Byles, Jeff. *Rubble: Unearthing the History of Demolition*. New York: Harmony Books, 2005.

Carlyle, Thomas. *Sartor Resartus*. New York: Oxford University Press, 1987.

——. "Signs of the Times." In *The Oxford Book of Essays*, edited by John Gross, 136–48. New York: Oxford University Press, 1991.

Carroll, James. *House of War: The Pentagon and the Disastrous Rise of American Power*. Boston: Houghton Mifflin, 2006.

Carter, Robert A. *Buffalo Bill Cody: The Man Behind the Legend*. Edison, NJ: Castle Books, 2000.

Carwardine, Richard. *Lincoln: A Life of Purpose and Power*. New York: Knopf, 2006.

Cather, Willa. *One of Ours*. New York: Vintage, 1971.

Catton, Bruce. *The Coming Fury*. London: Phoenix Press, 1961.

——. *Terrible Swift Sword*. London: Phoenix Press, 1963.

——. *Never Call Retreat*. London: Phoenix Press, 1965.

——. *Reflections on the Civil War*. Edited by John Leekley. New York: Promontory Press, 1998.

Catton, Bruce, and William Catton. *Two Roads to Sumter*. New York: McGraw-Hill, 1963.

Chaffin, Tom. *Pathfinder: John Charles Frémont and the Course of American Empire*. New York: Hill and Wang, 2002.

Chant, Christopher. *The History of North American Steam*. Edison, NJ: Chartwell Books, 2006.

Chartwell Books, ed. *From 1860 to the Present Day: The World's Great Machine Guns*. London: Chartwell Books, 1999.

Chen, Pauline. *Final Exam: A Surgeon's Reflections on Mortality*. New York: Alfred A. Knopf, 2007.

Chernow, Ron. *Titan: The Life of John D. Rockefeller, Sr.* New York: Vintage, 1999.

Christian, Shirley. *Before Lewis and Clark: The Story of the Chouteaus, the French Dynasty that Ruled America's Frontier*. New York: Farrar, Straus and Giroux, 2004.

Clifford, Frank. *The Backbone of the World: A Portrait of the Vanishing West Along the Continental Divide*. New York: Broadway Books, 2002.

Cody, W. F. *Buffalo Bill's Life Story: An Autobiography*. Mineola, NY: Dover Publications, 1920.

Cohen, Rachel. *A Chance Meeting: Intertwined Lives of American Writers and Artists 1854–1967*. New York: Random House, 2004.

Conant, Jennet. *109 East Palace: Robert Oppenheimer and the Secret City of Los Alamos*. New York: Simon & Schuster, 2005.

Connell, Evan S. *Son of the Morning Star: Custer and the Little Bighorn.* New York: Harper & Row, 1984.

Conrad, Joseph. *The Rescue: A Romance of the Shallows.* New York: Grosset & Dunlap, 1921.

Coombe, Jack D. *Gunfire Around the Gulf: The Last Major Naval Campaigns of the Civil War.* New York: Bantam Books, 1999.

Copping, Jasper. "Mystery of Great War's Lost Army Uncovered." *London Sunday Telegraph,* July 22, 2007, 14.

Crane, Stephen. *The Red Badge of Courage.* New York: Bantam, 1983.

Cronon, William. *Nature's Metropolis: Chicago and the Great West.* New York: W. W. Norton, 1991.

Davis, Richard Harding. *Captain Macklin: His Memoirs.* New York: Charles Scribner's Sons, 1917.

de Grazia, Victoria. *Irresistible Empire: America's Advance Through 20th-Century Europe.* Cambridge, MA: Harvard University Press, 2006.

DeLillo, Don. *Underworld.* New York: Simon & Schuster, 1997.

Denton, Sally. *Passion and Principle: John and Jesse Frémont, The Couple Whose Power, Politics, and Love Shaped Nineteenth-Century America.* New York: Bloomsbury, 2007.

DeVoto, Bernard. *The Year of Decision: 1846.* New York: Truman Talley Books, 2000.

——. *The Western Paradox: A Conservation Reader.* Edited by Douglas Brinkley and Patricia Nelson Limerick. New Haven, CT: Yale University Press, 2001.

Diamond, Jared. *Guns, Germs, and Steel: The Fates of Human Societies.* New York: W. W. Norton, 1999.

Dillard, Annie. *An Annie Dillard Reader.* New York: HarperPerennial, 1994.

DiLorenzo, Thomas J. *The Real Lincoln: A New Look at Abraham Lincoln, His Agenda, and an Unnecessary War.* New York: Three Rivers Press, 2002.

Dobyns, Kenneth W. *The Patent Office Pony: A History of the Early Patent Office.* Fredericksburg, VA: Sergeant Kirkland's Museum and Historical Society, 1994.

Doctorow, E. L. *The Waterworks.* New York: Signet, 1995.

——. *The March.* New York: Random House, 2005.

Donald, David Herbert. *Lincoln.* New York: Touchstone, 1995.

Donoghue, Denis. "The Pragmatic American: William James and Our Homegrown Way of Thought." *Harper's,* January 2007, 88–94.

Douglass, Frederick. *Narrative of the Life of Frederick Douglass, an American Slave, Written by Himself.* New York: Signet, 1968.

Einhorn, Robin L. *American Taxation, American Slavery.* Chicago: University of Chicago Press, 2006.

Eliot, George. *Middlemarch.* New York: New American Library, 1964.

Eliot. T.S. *The Sacred Wood: Essays on Poetry and Criticism.* London: Methune, 1920.

Ellis, John. *The Social History of the Machine Gun*. Baltimore: Johns Hopkins University Press, 1986.

Ely, Melvin Patrick. *Israel on the Appomattox: A Southern Experiment in Black Freedom from the 1790s Through the Civil War*. New York: Vintage, 2005.

Emerson, Ralph Waldo. *The Essential Writings of Ralph Waldo Emerson*. New York: Modern Library, 2000.

Faulks, Sebastian. *Birdsong*. London: Vintage, 1994.

Featherstone, Steve. "The Line is Hot: A History of the Machine Gun, Shot." *Harper's*, December 2005, 59–66.

Ferguson, Niall. *The Cash Nexus: Money and Power in the Modern World, 1700–2000*. New York: Basic Books, 2001.

——. *Colossus: The Rise and Fall of the American Empire*. New York: Penguin Books, 2005.

Foote, Shelby. *The Civil War: A Narrative: Fort Sumter to Perryville*. New York: Random House, 1958.

——. *The Civil War: A Narrative: Fredericksburg to Meridian*. New York: Random House, 1963.

——. *The Civil War: A Narrative: Red River to Appomattox*. New York: Random House, 1974.

——. *Stars in the Their Courses: The Gettysburg Campaign*. New York: Modern Library, 1994.

Ford, Roger. *The World's Great Machine Guns: From 1860 to the Present Day*. New York: Barnes & Noble, 1999.

Foster, Charles H. "The Modern Vulcan." *Potter's American Monthly* XII, no. 89 (May 1879): 321–33.

Freese, Barbara. *Coal: A Human History*. New York: Penguin Books, 2003.

Fromkin, David. *Europe's Last Summer: Who Started the Great War in 1914*? New York: Vintage, 2005.

Furgurson, Ernest B. *Freedom Rising: Washington in the Civil War*. New York: Vintage, 2005.

Gay, Peter. *The Naked Heart*. New York: W. W. Norton, 1995.

Giedion, Siegfried. *Mechanization Takes Command*. New York: W. W. Norton, 1969.

Gies, Frances and Joseph. *Cathedral, Forge, and Waterwheel: Technology and Invention in the Middle Ages*. New York: HarperCollins, 1994.

Goetzmann, William H. *New Lands, New Men: America and the Second Great Age of Discovery*. New York: Viking, 1986.

Golay, Michael. *Tide of Empire: America's March to the Pacific*. Hoboken, NJ: John Wiley and Sons, 2003.

Goodwin, Doris Kearns. *Team of Rivals: The Political Genius of Abraham Lincoln*. New York: Simon & Schuster, 2005.

Gordon, John Steele. *A Thread Across the Ocean: The Heroic Story of the Transatlantic Cable*. New York: Walker & Co., 2002.

——. *An Empire of Wealth: The Epic History of American Economic Power.* New York: Harper, 2004.

Gordon, Lyndall. *A Private Life of Henry James: Two Women and His Art.* New York: Norton, 1999.

Goudsblom, Johan. *Fire and Civilization.* New York: Penguin, 1994.

Grant, Ellsavorth S. "Gunmaker to the World." *American Heritage,* June 1968.

Grant, R. G. *Battle: A Visual Journey Through 5000 Years of Combat.* New York: DK Publishing, 2005.

Grant, Ulysses S. *Personal Memoirs of U. S. Grant.* Cambridge, MA: Da Capo Press, 2001.

Greve, Charles Theodore. *Centennial History of Cincinnati and Representative Citizens.* Ann Arbor: University of Michigan Press, 1904.

Griffith, Paddy. *Battle Tactics of the Civil War.* New Haven, CT: Yale University Press, 1989.

Guelzo, Allen G. *Lincoln's Emancipation Proclamation: The End of Slavery in America.* New York: Simon & Schuster, 2004.

Hagedorn, Ann. *Savage Peace: Hope and Fear in America, 1919.* New York: Simon & Schuster, 2007.

Hansen, Ron. *The Assassination of Jesse James by the Coward Robert Ford.* New York: Ballantine Books, 1983.

Harding, David, ed. *Weapons: An International Encyclopedia from 5000 B.C. to 2000 A.D.* New York: St. Martin's Press, 1990.

Harrison, Robert Pogue. *The Dominion of the Dead.* Chicago: University of Chicago Press, 2003.

Hays, Samuel P. *The Response to Industrialism, 1885–1914.* Chicago: University of Chicago Press, 1957.

Heaney, Seamus. *Beowulf.* New York: W. W. Norton & Co., 2000.

Henderson, G.F.R. *Stonewall Jackson & the American Civil War.* Old Saybrook, CT: Konecky & Konecky, 1897.

Henderson, Timothy J. *A Glorious Defeat: Mexico and Its War with the United States.* New York: Hill and Wang, 2007.

Henkin, David M. *The Postal Age: The Emergence of Modern Communications in Nineteenth-Century America.* Chicago: University of Chicago Press, 2006.

Himmelfarb, Gertrude. *The Roads to Modernity: The British, French, and American Enlightenments.* New York: Vintage, 2005.

Hoffman, Ian. "Lab Unveils Gun to Scare Off Terrorists." *Oakland Tribune,* February 3, 2006, 1.

Holmes, Oliver Wendell. *The Autocrat of the Breakfast-Table.* New York: Airmont, 1968.

Holmes, Richard. *Wellington: The Iron Duke.* London: HarperCollins, 2003.

Hosley, William. *Colt: The Making of an American Legend.* Amherst: University of Massachusetts Press, 1996.

Howells, William Dean. *The Rise of Silas Lapham.* New York: W. W. Norton, 1982.

Hughes, James B. *The Gatling Gun Notebook: A Collection of Data and Illustrations.* Lincoln, RI: Andrew Mowbray, 2000.

Hughes, Thomas P. *The Human-Built World: How to Think About Technology and Culture.* Chicago: University of Chicago Press, 2005.

Hunter, Louis C. *Steamboats on the Western Rivers: An Economic and Technological History.* New York: Dover, 1993.

Hunter, Stephen. "Out with a Bang: The Loss of the Classic Winchester is Loaded with Symbolism." *Washington Post,* June 30, 2006, C-1.

Isaacson, Walter. *Einstein: His Life and Universe.* New York: Simon & Schuster, 2007.

James, Henry. *The Golden Bowl.* New York: Penguin, 2001.

James, William. *Memories and Studies.* Cambridge, MA: Riverside Press, 1911.

Johnson, F. Roy, and E. Frank Stephenson, Jr. *The Gatling Gun and the Flying Machine.* Murfreesboro, NC: Johnson Publishing, 1979.

Johnson, Paul. *The Birth of the Modern: World Society 1815–1830.* New York: HarperCollins, 1991.

Jones, Howard Mumford. *The Age of Energy: Varieties of American Experience, 1865–1915.* New York: Viking, 1970.

Jones, Jill. *Empires of Light: Edison, Tesla, Westinghouse, and the Race to Electrify the World.* New York: Random House, 2003.

Jones, Landon Y. *William Clark and the Shaping of the West.* New York: Hill and Wang, 2004.

Jones, Russell Celyn. *Ten Seconds from the Sun.* London: Abacus, 2006.

Kahaner, Larry. "Weapon of Mass Destruction." *Washington Post,* November 26, 2006, B1.

——. *AK-47: The Weapon That Changed the Face of War.* Hoboken, NJ: John Wiley and Sons, 2007.

Kammen, Michael. *Mystic Chords of Memory: The Transformation of Tradition in American Culture.* New York: Vintage, 1993.

Kaplan, Fred. *The Singular Mark Twain.* New York: Anchor Books, 2005.

Kasper, Shirl. *Annie Oakley.* Norman, OK: University of Oklahoma Press, 1992.

Kauffman, Michael W. *American Brutus: John Wilkes Booth and the Lincoln Conspiracies.* New York: Random House, 2004.

Keegan, John. *The Face of Battle.* New York: Penguin, 1976.

——, ed. *The Penguin Book of War: Great Military Writing.* New York: Viking, 1999.

Kelly, Jack. *Gunpowder: Alchemy, Bombards, and Pyrotechnics: The History of the Explosive that Changed the World.* New York: Basic Books, 2004.

Kennedy, Paul. *The Rise and Fall of the Great Powers.* New York: Random House, 1987.

——. "The Worst of Times?" *New York Review of Books*, November 2, 2006, 23–25.

Kessner, Thomas. *Capital City: New York City and the Men Behind America's Rise to Economic Dominance, 1860–1900*. New York: Simon & Schuster, 2003.

Khan, B. Zorina. *The Democratization of Invention: Patents and Copyrights in American Economic Development, 1790–1920*. Cambridge: Cambridge University Press, 2005.

Konstam, Angus. *The Pocket Book of Civil War Weapons*. London: Greenwich, 2004.

Kurlansky, Mark. *Cod*. New York: Penguin, 1997.

Langguth, A. J. *Union 1812: The Americans Who Fought the Second War of Independence*. New York: Simon & Schuster, 2006.

Larson, Erik. *The Devil in the White City*. New York: Vintage, 2003.

Leach, William. *Land of Desire: Merchants, Power, and the Rise of a New American Culture*. New York: Vintage, 1994.

Lepore, Jill. *A Is for American: Letters and Other Characters in the Newly United States*. New York: Vintage, 2003.

Lindsey, Almont. *The Pullman Strike: The Story of a Unique Experiment and of a Great Labor Upheaval*. Chicago: University of Chicago Press, 1942.

Lord, Walter. *The Good Years: From 1900 to the First World War*. New York: Harper & Brothers, 1960.

Lubetkin, M. John. *Jay Cooke's Gamble: The Northern Pacific Railroad, the Sioux, and the Panic of 1873*. Norman, OK: University of Oklahoma Press, 2006.

Lynn, John A. *Tools of War: Instruments, Ideas, and Institutions of Warfare, 1445–1871*. Urbana: University of Illinois Press, 1990.

MacDonald, Ann-Marie. *Fall On Your Knees*. Toronto: Vintage, 1997.

McFarland, Philip. *Hawthorne in Concord*. New York: Grove Press, 2004.

McGahern, John. *The Dark*. New York: Penguin Books, 1983.

McKendrik, Neil. Introduction to *The Vickers Brothers: Armaments and Enterprise, 1854–1924*, by Clive Trebilcock. London: Europa Publications, 1977.

McMurtry, Larry. *Sin Killer*. New York: Simon & Schuster, 2002.

——. *By Sorrow's River*. New York: Simon & Schuster, 2003.

——. *The Wandering Hill*. New York: Simon & Schuster, 2003.

——. *The Colonel and Little Missie: Buffalo Bill, Annie Oakley, and the Beginnings of Superstardom in America*. New York: Simon & Schuster, 2005.

McNab, Chris. *Twentieth-Century Small Arms*. London: Amber Books, 2001.

——. *Gun: A Visual History*. New York: DK Publishing, 2007.

McNeill, William. *The Pursuit of Power*. Chicago: University of Chicago Press, 1982.

McPherson, James M. *Battle Cry of Freedom: The Civil War Era*. New York: Ballantine Books, 1988.

Madden, W. C. *Crown Hill Cemetery*. Charleston, SC: Arcadia Publishing, 2004.

Manchester, William. *The Arms of Krupp: 1587–1968*. New York: Bantam, 1970.

Mazrim, Robert. *The Sangamo Frontier: History & Archaeology in the Shadow of Lincoln.* Chicago: University of Chicago Press, 2007.

Melville, Herman. *White Jacket; or, the World in a Man-of-War.* New York: New American Library, 1979.

Menand, Louis. *The Metaphysical Club: A Story of Ideas in America.* New York: Farrar, Straus and Giroux, 2001.

Merk, Frederick. *Manifest Destiny and Mission in American History.* New York: Vintage, 1966.

Meron, Thomas. *Henry's Wars and Shakespeare's Laws: Perspectives on the Law of War in the Late Middle Ages.* Oxford: Clarendon Press, 1993.

Meyer, G. J. *A World Undone: The Story of the Great War, 1914 to 1918.* New York: Delacorte, 2006.

Miller, William Lee. *Lincoln's Virtues: An Ethical Biography.* New York: Vintage, 2003.

Murray, Stuart. *Atlas of American Military History.* New York: Checkmark Books, 1995.

Negri, Paul, ed. *Civil War Poetry: An Anthology.* New York: Dover, 1997.

Nelson, Scott Reynolds. *Steel Drivin' Man: John Henry: The Untold Story of an American Legend.* New York: Oxford University Press, 2006.

Newbolt, Sir Henry. "Vitai Lampada." In *The Oxford Book of War Poetry,* edited by Jon Stallworthy, 243. New York: Oxford University Press, 1984.

Nye, David. *American Technological Sublime.* Cambridge, MA: MIT Press, 1994.

——. *America as Second Creation: Technology and Narratives of New Beginning.* Cambridge, MA: MIT Press, 2003.

O'Connell, Robert L. *Soul of the Sword: An Illustrated History of Weaponry and Warfare from Prehistory to the Present.* New York: Free Press, 2002.

Olds, Bruce. *Raising Holy Hell.* New York: Henry Holt, 1995.

Oppenheimer, Robert. "Robert Oppenheimer on Albert Einstein." In *The Company They Kept: Writers on Unforgettable Friendships,* edited by Robert B. Silvers and Barbara Epstein, 35–41. New York: New York Review Books, 2006.

Parker, Geoffrey. *The Military Revolution: Military Innovation and the Rise of the West, 1500–1800.* London: Cambridge University Press, 1998.

Parker, John H. *Tactical Organization and Uses of Machine Guns in the Field.* Kansas City, MO: Hudson-Kimberly Publishing, 1899.

Parramore, Thomas C. "The North Carolina Background of Richard Jordan Gatling." *North Carolina Historical Review,* January 1964, 54–60.

Patterson, James T. *Restless Giant: The United States from Watergate to Bush v. Gore.* New York: Oxford University Press, 2005.

Pearl, Matthew. *The Dante Club.* New York: Ballantine Books, 2006.

Pelecanos, George P. *The Sweet Forever.* New York: Dell, 1998.

Phillips, Kevin. *American Theocracy: The Peril and Politics of Radical Religion, Oil, and Borrowed Money in the 21st Century.* New York: Viking Penguin, 2006.

Postgate, Raymond. *The Story of a Year: 1848.* London: Cassell, 1955.
Powers, Ron. *Mark Twain: A Life.* New York: Free Press, 2005.
Preston, Richard. *The Demon in the Freezer.* New York: Random House, 2002.
Price, Anthony. *The "Old Vengeful."* London: Orion Books, 2003.
Priest, Dana. *The Mission: Waging War and Keeping Peace with America's Military.* New York: W. W. Norton, 2003.
Quammen, David. *The Reluctant Mr. Darwin: An Intimate Portrait of Charles Darwin and the Making of His Theory of Evolution.* New York: W. W. Norton, 2006.
Raban, Jonathan. *Old Glory: An American Voyage.* New York: Simon & Schuster, 1981.
Ray, Angela G. *The Lyceum and Public Culture in the Nineteenth-Century United States.* East Lansing, MI: Michigan State University Press, 2005.
Ray, William and Marlys. *The Art of Invention.* Princeton, NJ: Pyne Press, 1974.
Reback, Gary L. "Patently Absurd: Too Many Patents Are Just as Bad for Society as Too Few." *Forbes,* June 24, 2000, 44.
Reece, Erik. *Lost Mountain.* New York: Riverhead Books, 2007.
Remarque, Erich Maria. *All Quiet on the Western Front.* New York: Fawcett Crest, 1958.
Remini, Robert. *Daniel Webster: The Man and His Time.* New York: W. W. Norton, 1997.
Riley, Glenda. *The Life and Legacy of Annie Oakley.* Norman, OK: University of Oklahoma Press, 1994.
Robertson, James L., Jr., ed. *Stonewall Jackson's Book of Maxims.* Nashville: Cumberland House, 2002.
Roosevelt, Theodore. *The Rough Riders.* New York: Barnes & Noble, 2004.
Rosen, Fred. *Gold! The Story of the 1848 Gold Rush and How It Shaped a Nation.* New York: Thunder's Mouth Press, 2005.
Russell, Charles Edward. *A-Rafting on the Mississip'.* Minneapolis: University of Minnesota Press, 2001.
Rybczynski, Witold. *A Clearing in the Distance: Frederick Law Olmsted and America in the 19th Century.* New York: Touchstone, 2000.
Rydell, Robert W., and Bob Kroes. *Buffalo Bill in Bologna: The Americanization of the World, 1869–1922.* Chicago: University of Chicago Press, 2005.
Sachs, Aaron. *The Humboldt Current: Nineteenth-Century Exploration and the Roots of American Environmentalism.* New York: Viking, 2006.
Sale, Kirkpatrick. *The Fire of His Genius: Robert Fulton and the American Dream.* New York: Touchstone, 2001.
Schama, Simon. *Dead Certainties (Unwarranted Speculations).* New York: Alfred A. Knopf, 1991.
Sears, Stephen W. *Landscape Turned Red: The Battle of Antietam.* New York: Ticknor & Fields, 1983.
Shaara, Michael. *The Killer Angels.* New York: Ballantine Books, 1975.
Shaw, George Bernard. *Saint Joan, Major Barbara, Androcles the Lion.* New York: Modern Library, 1956.

Shaw, Robert B. *History of the Comstock Patent Medicine Business and Dr. Morse's Indian Root Pills.* Washington, D.C.: Smithsonian Institution Press, 1972.

Silverman, Kenneth. *Lightning Man: The Accursed Life of Samuel F. B. Morse.* Cambridge, MA: Da Capo Press, 2003.

Slaughter, Thomas P. *Exploring Lewis and Clark: Reflections on Man and Wilderness.* New York: Vintage, 2004.

Slotkin, Richard. *The Fatal Environment: The Myth of the Frontier in the Age of Industrialization, 1800–1890.* New York: HarperPerennial, 1994.

Smith, Anthony. *Machine Gun: The Story of the Men and the Weapon That Changed the Face of War.* London: Piatkus, 2002.

Smith, Jean Edward. *Grant.* New York: Simon & Schuster, 2001.

Smith, Phillip. *Why War? The Cultural Logic of Iraq, the Gulf War, and Suez.* Chicago: University of Chicago Press, 2006.

Spragg, Mark. *Where Rivers Change Direction.* New York: Riverhead Books, 1999.

Steeples, Douglas and David O. Whitten. *Democracy in Desperation: The Depression of 1893.* Westport, CT: Greenwood Press, 1998.

Stephenson, E. Frank, Jr. *Gatling: A Photographic Remembrance.* Murfreesboro, NC: Meherrin River Press, 1993.

Stevenson, David. *Cataclysm: The First World War as Political Tragedy.* New York: Basic Books, 2005.

Stiles, T. J. *Jesse James: Last Rebel of the Civil War.* New York: Vintage, 2003.

Stout, Harry S. *Upon the Altar of the Nation: A Moral History of the American Civil War.* New York: Viking, 2006.

Stowe, Harriet Beecher. *Uncle Tom's Cabin.* New York: Barnes & Noble, 2003.

Strachey, Lytton. *Eminent Victorians.* New York: Penguin, 1986.

Stross, Randall. *The Wizard of Menlo Park: How Thomas Alva Edison Invented the Modern World.* New York: Crown, 2007.

Strouse, Jean. *Alice James: A Biography.* Boston: Houghton Mifflin, 1980.

——. *Morgan: American Financier.* New York: HarperCollins, 1999.

Styron, William. *The Confessions of Nat Turner.* New York: Modern Library, 1994.

Swift, Jonathan. *Gulliver's Travels and Other Writings.* New York: Modern Library, 1958.

Symonds, A.J.A. *The Quest for Corvo.* Baltimore: Penguin, 1966.

Tankersley, Jim, and Joshua Boak. "Business as Usual." *Toledo Blade*, September 17, September 24 and October 1, 2006.

Thoreau, Henry David. *The Writing of Henry David Thoreau,* Vol. 5. Boston: Houghton Mifflin, 1906.

Tillyard, Stella. "All Our Pasts: The Rise of Popular History." *Times Literary Supplement*, October 13, 2006, 7–9.

Timmons, Todd. *Science and Technology in Nineteenth-Century America.* Westport, CT: Greenwood Press, 2005.

Tomalin, Claire. *Thomas Hardy: The Time-Torn Man.* New York: Penguin Press, 2007.

Toobin, Jeffrey. "Google's Moon Shot: The Quest for the Universal Library." *New Yorker,* February 5, 2007, 30–35.

Torgovnick, Marianna. *The War Complex: World War II in Our Time.* Chicago: University of Chicago Press, 2005.

Townshend, Charles, ed. *The Oxford History of Modern War.* Oxford: Oxford University Press, 2005.

Trachtenberg, Alan. *The Incorporation of America: Culture and Society in the Gilded Age.* New York: Hill and Wang, 2007.

Traxel, David. *1898: The Birth of the American Century.* New York: Vintage, 1999.

Trebilcock, Clive. *The Vickers Brothers: Armaments and Enterprise, 1854–1914.* London: Europa Publications, 1977.

Tsouras, Peter. *Gettysburg: An Alternate History.* London: Greenhill Books, 1997.

Tuchman, Barbara. *The Guns of August.* New York: Dell, 1962.

Turner, Frederick Jackson. "The Significance of the Frontier in American History." In *The American Intellectual Tradition,* edited by David Hollinger and Charles Capper, 84–92. New York: Oxford University Press, 2001.

Twain, Mark. *A Connecticut Yankee in King Arthur's Court.* New York: Bantam, 1981.

——. *Adventures of Tom Sawyer.* New York: Barnes & Noble, 2006.

Uglow, Jenny. *The Lunar Men: Five Friends Whose Curiosity Changed the World.* New York: Farrar, Straus and Giroux, 2002.

Van Creveld, Martin. "Technology and War I." In *The Oxford History of Modern War,* edited by Charles Townshend, 201–23. New York: Oxford University Press, 2005.

Vidal, Gore. *1876: A Novel.* New York: Random House, 1976.

Vollmann, William T. *Europe Central.* New York: Viking, 2005.

Wahl, Paul, and Don Toppel. *The Gatling Gun.* New York: Arco Publishing, 1965.

Wall, Joseph Frazier. *Andrew Carnegie.* Pittsburgh: University of Pittsburgh Press, 1989.

Ward, Geoffrey C., Dayton Duncan, and Ken Burns. *Mark Twain.* New York: Alfred A. Knopf, 2001.

Watts, Sarah. *Rough Rider in the White House: Theodore Roosevelt and the Politics of Desire.* Chicago: University of Chicago Press, 2006.

Weber, Jennifer L. *Copperheads: The Rise and Fall of Lincoln's Opponents in the North.* New York: Oxford University Press, 2006.

Wert, Jeffrey D. *The Sword of Lincoln: The Army of the Potomac.* New York: Simon & Schuster, 2005.

Wheeler, Tom. *Mr. Lincoln's T-Mails: The Untold Story of How Abraham Lincoln Used the Telegraph to Win the Civil War.* New York: HarperCollins, 2006.

Whipple, Bishop Henry. *Bishop Whipple's Southern Diary 1843–1844.* Edited by Lester B. Shippee. Minneapolis: University of Minnesota Press, 1937.

Whitman, Walt. *Walt Whitman's Civil War.* Edited by Walter Lowenfels. New York: Da Capo Press, 1960.

Wiebe, Robert H. *The Search for Order.* New York: Hill and Wang, 1967.

Wills, Garry. *Henry Adams and the Making of America.* Boston: Houghton Mifflin, 2005.

Wilson, Edmund. *Patriotic Gore: Studies in the Literature of the American Civil War.* New York: W. W. Norton, 1994.

Winik, Jay. *April 1865.* New York: HarperCollins, 2001.

Wright, Robert E. *The First Wall Street: Chesnut Street, Philadelphia, & the Birth of American Finance.* Chicago: University of Chicago Press, 2005.

Wright, Robert E., and David J. Cowen. *Financial Founding Fathers: Men Who Made America Rich.* Chicago: University of Chicago Press, 2006.

Zelazny, Roger. *Doorways in the Sand.* New York: Harper & Row, 1976.

INDEX